Human Body Systems and Health

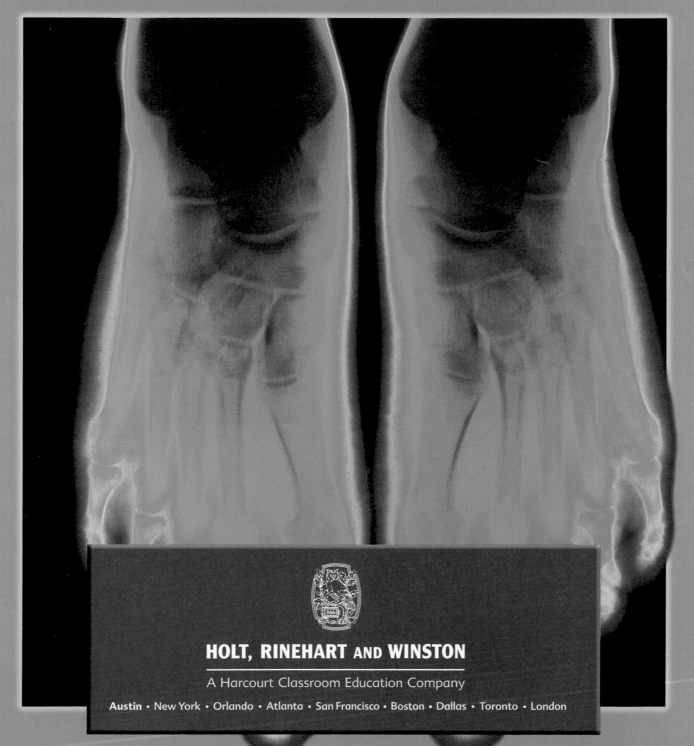

HOLT, RINEHART AND WINSTON

A Harcourt Classroom Education Company

Austin · New York · Orlando · Atlanta · San Francisco · Boston · Dallas · Toronto · London

Acknowledgments

Chapter Writers

Katy Z. Allen
Science Writer and Former Biology Teacher
Wayland, Massachusetts

Linda Ruth Berg, Ph.D.
Adjunct Professor–Natural Sciences
St. Petersburg Junior College
St. Petersburg, Florida

Jennie Dusheck
Science Writer
Santa Cruz, California

Mark F. Taylor, Ph.D.
Associate Professor of Biology
Baylor University
Waco, Texas

Lab Writers

Diana Scheidle Bartos
Science Consultant and Educator
Diana Scheidle Bartos, L.L.C.
Lakewood, Colorado

Carl Benson
General Science Teacher
Plains High School
Plains, Montana

Charlotte Blassingame
Technology Coordinator
White Station Middle School
Memphis, Tennessee

Marsha Carver
Science Teacher and Dept. Chair
McLean County High School
Calhoun, Kentucky

Kenneth E. Creese
Science Teacher
White Mountain Junior High School
Rock Springs, Wyoming

Linda Culp
Science Teacher and Dept. Chair
Thorndale High School
Thorndale, Texas

James Deaver
Science Teacher and Dept. Chair
West Point High School
West Point, Nebraska

Frank McKinney, Ph.D.
Professor of Geology
Appalachian State University
Boone, North Carolina

Alyson Mike
Science Teacher
East Valley Middle School
East Helena, Montana

C. Ford Morishita
Biology Teacher
Clackamas High School
Milwaukie, Oregon

Patricia D. Morrell, Ph.D.
Assistant Professor, School of Education
University of Portland
Portland, Oregon

Hilary C. Olson, Ph.D.
Research Associate
Institute for Geophysics
The University of Texas
Austin, Texas

James B. Pulley
Science Editor and Former Science Teacher
Liberty High School
Liberty, Missouri

Denice Lee Sandefur
Science Chairperson
Nucla High School
Nucla, Colorado

Patti Soderberg
Science Writer
The BioQUEST Curriculum Consortium
Beloit College
Beloit, Wisconsin

Phillip Vavala
Science Teacher and Dept. Chair
Salesianum School
Wilmington, Delaware

Albert C. Wartski
Biology Teacher
Chapel Hill High School
Chapel Hill, North Carolina

Lynn Marie Wartski
Science Writer and Former Science Teacher
Hillsborough, North Carolina

Ivora D. Washington
Science Teacher and Dept. Chair
Hyattsville Middle School
Washington, D.C.

Academic Reviewers

Renato J. Aguilera, Ph.D.
Associate Professor
Department of Molecular, Cell, and Developmental Biology
University of California
Los Angeles, California

David M. Armstrong, Ph.D.
Professor of Biology
Department of E.P.O. Biology
University of Colorado
Boulder, Colorado

Alissa Arp, Ph.D.
Director and Professor of Environmental Studies
Romberg Tiburon Center
San Francisco State University
Tiburon, California

Russell M. Brengelman
Professor of Physics
Morehead State University
Morehead, Kentucky

John A. Brockhaus, Ph.D.
Director of Mapping, Charting, and Geodesy Program
Department of Geography and Environmental Engineering
United States Military Academy
West Point, New York

Linda K. Butler, Ph.D.
Lecturer of Biological Sciences
The University of Texas
Austin, Texas

Barry Chernoff, Ph.D.
Associate Curator
Division of Fishes
The Field Museum of Natural History
Chicago, Illinois

Donna Greenwood Crenshaw, Ph.D.
Instructor
Department of Biology
Duke University
Durham, North Carolina

Hugh Crenshaw, Ph.D.
Assistant Professor of Zoology
Duke University
Durham, North Carolina

Joe W. Crim, Ph.D.
Professor of Biology
University of Georgia
Athens, Georgia

Peter Demmin, Ed.D.
Former Science Teacher and Chair
Amherst Central High School
Amherst, New York

Joseph L. Graves, Jr., Ph.D.
Associate Professor of Evolutionary Biology
Arizona State University West
Phoenix, Arizona

William B. Guggino, Ph.D.
Professor of Physiology and Pediatrics
The Johns Hopkins University School of Medicine
Baltimore, Maryland

David Haig, Ph.D.
Assistant Professor of Biology
Department of Organismic and Evolutionary Biology
Harvard University
Cambridge, Massachusetts

Roy W. Hann, Jr., Ph.D.
Professor of Civil Engineering
Texas A&M University
College Station, Texas

Printed in the United States of America

ISBN 0-03-064779-7

1 2 3 4 5 6 7   048   05 04 03 02 01 00

Acknowledgments (cont.)

John E. Hoover, Ph.D.
Associate Professor of Biology
Millersville University
Millersville, Pennsylvania

Joan E. N. Hudson, Ph.D.
Associate Professor of Biological Sciences
Sam Houston State University
Huntsville, Texas

Laurie Jackson-Grusby, Ph.D.
Research Scientist and Doctoral Associate
Whitehead Institute for Biomedical Research
Massachusetts Institute of Technology
Cambridge, Massachusetts

George M. Langford, Ph.D.
Professor of Biological Sciences
Dartmouth College
Hanover, New Hampshire

Melanie C. Lewis, Ph.D.
Professor of Biology, Retired
Southwest Texas State University
San Marcos, Texas

V. Patteson Lombardi, Ph.D.
Research Assistant Professor of Biology
Department of Biology
University of Oregon
Eugene, Oregon

Glen Longley, Ph.D.
Professor of Biology and Director of the Edwards Aquifer Research Center
Southwest Texas State University
San Marcos, Texas

William F. McComas, Ph.D.
Director of the Center to Advance Science Education
University of Southern California
Los Angeles, California

LaMoine L. Motz, Ph.D.
Coordinator of Science Education
Oakland County Schools
Waterford, Michigan

Nancy Parker, Ph.D.
Associate Professor of Biology
Southern Illinois University
Edwardsville, Illinois

Barron S. Rector, Ph.D.
Associate Professor and Extension Range Specialist
Texas Agricultural Extension Service
Texas A&M University
College Station, Texas

Peter Sheridan, Ph.D.
Professor of Chemistry
Colgate University
Hamilton, New York

Miles R. Silman, Ph.D.
Assistant Professor of Biology
Wake Forest University
Winston-Salem, North Carolina

Neil Simister, Ph.D.
Associate Professor of Biology
Department of Life Sciences
Brandeis University
Waltham, Massachusetts

Lee Smith, Ph.D.
Curriculum Writer
MDL Information Systems, Inc.
San Leandro, California

Robert G. Steen, Ph.D.
Manager, Rat Genome Project
Whitehead Institute—Center for Genome Research
Massachusetts Institute of Technology
Cambridge, Massachusetts

Martin VanDyke, Ph.D.
Professor of Chemistry, Emeritus
Front Range Community College
Westminister, Colorado

E. Peter Volpe, Ph.D.
Professor of Medical Genetics
Mercer University School of Medicine
Macon, Georgia

Harold K. Voris, Ph.D.
Curator and Head
Division of Amphibians and Reptiles
The Field Museum of Natural History
Chicago, Illinois

Mollie Walton
Biology Instructor
El Paso Community College
El Paso, Texas

Peter Wetherwax, Ph.D.
Professor of Biology
University of Oregon
Eugene, Oregon

Mary K. Wicksten, Ph.D.
Professor of Biology
Texas A&M University
College Station, Texas

R. Stimson Wilcox, Ph.D.
Associate Professor of Biology
Department of Biological Sciences
Binghamton University
Binghamton, New York

Conrad M. Zapanta, Ph.D.
Research Engineer
Sulzer Carbomedics, Inc.
Austin, Texas

Safety Reviewer

Jack Gerlovich, Ph.D.
Associate Professor
School of Education
Drake University
Des Moines, Iowa

Teacher Reviewers

Barry L. Bishop
Science Teacher and Dept. Chair
San Rafael Junior High School
Ferron, Utah

Carol A. Bornhorst
Science Teacher and Dept. Chair
Bonita Vista Middle School
Chula Vista, California

Paul Boyle
Science Teacher
Perry Heights Middle School
Evansville, Indiana

Yvonne Brannum
Science Teacher and Dept. Chair
Hine Junior High School
Washington, D.C.

Gladys Cherniak
Science Teacher
St. Paul's Episcopal School
Mobile, Alabama

James Chin
Science Teacher
Frank A. Day Middle School
Newtonville, Massachusetts

Kenneth Creese
Science Teacher
White Mountain Junior High School
Rock Springs, Wyoming

Linda A. Culp
Science Teacher and Dept. Chair
Thorndale High School
Thorndale, Texas

Georgiann Delgadillo
Science Teacher
East Valley Continuous Curriculum School
Spokane, Washington

Alonda Droege
Biology Teacher
Evergreen High School
Seattle, Washington

Michael J. DuPré
Curriculum Specialist
Rush Henrietta Junior-Senior High School
Henrietta, New York

Rebecca Ferguson
Science Teacher
North Ridge Middle School
North Richland Hills, Texas

Susan Gorman
Science Teacher
North Ridge Middle School
North Richland Hills, Texas

Gary Habeeb
Science Mentor
Sierra-Plumas Joint Unified School District
Downieville, California

Karma Houston-Hughes
Science Mentor
Kyrene Middle School
Tempe, Arizona

Roberta Jacobowitz
Science Teacher
C. W. Otto Middle School
Lansing, Michigan

Kerry A. Johnson
Science Teacher
Isbell Middle School
Santa Paula, California

M. R. Penny Kisiah
Science Teacher and Dept. Chair
Fairview Middle School
Tallahassee, Florida

Kathy LaRoe
Science Teacher
East Valley Middle School
East Helena, Montana

Jane M. Lemons
Science Teacher
Western Rockingham Middle School
Madison, North Carolina

Scott Mandel, Ph.D.
Director and Educational Consultant
Teachers Helping Teachers
Los Angeles, California

Thomas Manerchia
Former Biology and Life Science Teacher
Archmere Academy
Claymont, Delaware

Maurine O. Marchani
Science Teacher and Dept. Chair
Raymond Park Middle School
Indianapolis, Indiana

Jason P. Marsh
Biology Teacher
Montevideo High School and Montevideo Country School
Montevideo, Minnesota

Edith C. McAlanis
Science Teacher and Dept. Chair
Socorro Middle School
El Paso, Texas

Kevin McCurdy, Ph.D.
Science Teacher
Elmwood Junior High School
Rogers, Arkansas

Kathy McKee
Science Teacher
Hoyt Middle School
Des Moines, Iowa

Acknowledgments continue on page 211.

Contents **iii**

D Human Body Systems and Health

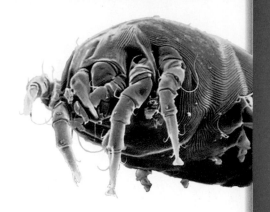

Skills Development

Process Skills

QuickLabs

Chapter Labs

Skills Development (continued)

Research and Critical Thinking Skills

Apply

Feature Articles

Science, Technology, and Society

Eureka!

Weird Science

Health Watch

Across the Sciences

Careers

Connections

To the Student

This book was created to make your science experience interesting, exciting, and fun!

Go for It!

Science is a process of discovery, a trek into the unknown. The skills you develop using *Holt Science & Technology*— such as observing, experimenting, and explaining observations and ideas— are the skills you will need for the future. There is a universe of exploration and discovery awaiting those who accept the challenges of science.

Science & Technology

You see the interaction between science and technology every day. Science makes technology possible. On the other hand, some of the products of technology, such as computers, are used to make further scientific discoveries. In fact, much of the scientific work that is done today has become so technically complicated and expensive that no one person can do it entirely alone. But make no mistake, the creative ideas for even the most highly technical and expensive scientific work still come from individuals.

Activities and Labs

The activities and labs in this book will allow you to make some basic but important scientific discoveries on your own. You can even do some exploring on your own at home! Here's your chance to use your imagination and curiosity as you investigate your world.

Keep a ScienceLog

In this book, you will be asked to keep a type of journal called a ScienceLog to record your thoughts, observations, experiments, and conclusions. As you develop your ScienceLog, you will see your own ideas taking shape over time. You'll have a written record of how your ideas have changed as you learn about and explore interesting topics in science.

Know "What You'll Do"

The "What You'll Do" list at the beginning of each section is your built-in guide to what you need to learn in each chapter. When you can answer the questions in the Section Review and Chapter Review, you know you are ready for a test.

Check Out the Internet

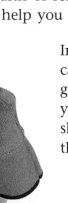

You will see this ᔊᴸᴵᴺᴷˢ logo throughout the book. You'll be using sciLINKS as your gateway to the Internet. Once you log on to sciLINKS using your computer's Internet link, type in the sciLINKS address. When asked for the keyword code, type in the keyword for that topic. A wealth of resources is now at your disposal to help you learn more about that topic.

In addition to sciLINKS you can log on to some other great resources to go with your text. The addresses shown below will take you to the home page of each site.

internet connect

This textbook contains the following on-line resources to help you make the most of your science experience.

 go. hrw .com

Visit **go.hrw.com** for extra help and study aids matched to your textbook. Just type in the keyword HST HOME.

 SCiLINKS NSTA

Visit **www.scilinks.org** to find resources specific to topics in your textbook. Keywords appear throughout your book to take you further.

 Smithsonian Institution® Internet Connections

Visit **www.si.edu/hrw** for specifically chosen on-line materials from one of our nation's premier science museums.

 **CNNfyi.com**

Visit **www.cnnfyi.com** for late-breaking news and current events stories selected just for you.

Body Organization and Structure

Sections

Pre-Reading
Questions

1. What is the relationship between cells, tissues, and organs?

2. How do your skin, muscles, and bones help to keep you well?

A Winning Effort

In 1999 and 2000, Lance Armstrong won the Tour de France. This victory was amazing because a few years earlier, Lance was diagnosed with cancer, which had weakened him. But with medicine and hard work, Lance's body grew strong again—strong enough for him to win one of the toughest races in all of sports. In this chapter, you will learn more about how each of the organ systems of the human body works to keep you healthy.

TOO COLD FOR COMFORT

Did you know that your nervous system sends you messages about your body's cells? For example, the pain you feel when someone steps on your toe is a message that you should move your toe to safety. Try this exercise to watch your nervous system in action.

Procedure

1. Hold a **few pieces of ice** in one hand. Allow the melting water to drip into a **dish.** Hold the ice until the cold is uncomfortable. Then release the ice into the dish. What message did you receive from your nervous system?

2. Look at the hand that held the ice, and then look at your other hand. What changes in your skin do you see? How quickly does the cold hand return to normal?

Analysis

3. What organ systems do you think were involved in restoring your hand to normal?

4. Think of a time when your nervous system sent you a message, such as an uncomfortable feeling of heat, cold, or pain. How did your body react? Which organ systems do you think were involved in the reaction?

Body Organization

Terms to Learn

homeostasis
tissue
epithelial tissue
nervous tissue

muscle tissue
connective tissue
organ

What You'll Do

♦ Identify the major tissues found in the body.
♦ Compare an organ with an organ system.
♦ Describe a major function of each organ system.

Your body has an amazing ability to survive, even in the face of harsh conditions. How does a person stay alive even though the environment around him or her is so cold? A short answer is that the body did not allow its internal conditions to change enough to stop the cells from working properly. The maintenance of a stable internal environment is called **homeostasis** (HOH mee oh STAY sis). If homeostasis is disrupted, cells suffer and sometimes die.

Four Types of Tissue

Making sure your internal environment remains stable enough to support healthy cells is not an easy task. Many different "jobs" must be done to maintain homeostasis. Fortunately, not every cell has to do all those jobs because the cells are organized into different teams. Just as each member of a soccer team has a special role in the game, each cell in your body has a specific job in maintaining homeostasis. A group of similar cells working together forms a **tissue**. Your body contains four main types of tissue—epithelial tissue, connective tissue, muscle tissue, and nervous tissue, as shown in **Figure 1.**

Figure 1 *Your body has four types of tissue, and each type has a special function in your body.*

Epithelial tissue covers and protects underlying tissue. When you look at the surface of your skin, you see epithelial tissue. The cells stick tightly and form a continuous sheet.

Nervous tissue sends electrical signals through the body. It is found in the brain, nerves, and sense organs.

Tissues Form Organs

Two or more tissues working together form an **organ.** One type of tissue alone cannot do all the things that several types working together can do. Your stomach, as shown in **Figure 2,** uses several different types of tissue to carry out digestion.

Organs Form Systems

Your stomach does much to help you digest your food, but it doesn't do it all. It works together with other organs, such as the small intestine and large intestine, to digest your food. Organs working together make up an *organ system.* The failure of any part can affect the entire system. Your body has 11 major organ systems, which are illustrated on the next two pages. Are there any that you have not heard of before?

The Stomach Is an Organ

Nervous tissue in the stomach partly controls the production of acids that aid in the digestion of food. Nervous tissue signals when the stomach is full.

Epithelial tissue lines the stomach.

Layers of **muscle tissue** break up stomach contents.

Blood and a **connective tissue** called collagen are found in the wall of the stomach.

Figure 2 *The four types of tissue work together so that the stomach can carry out digestion.*

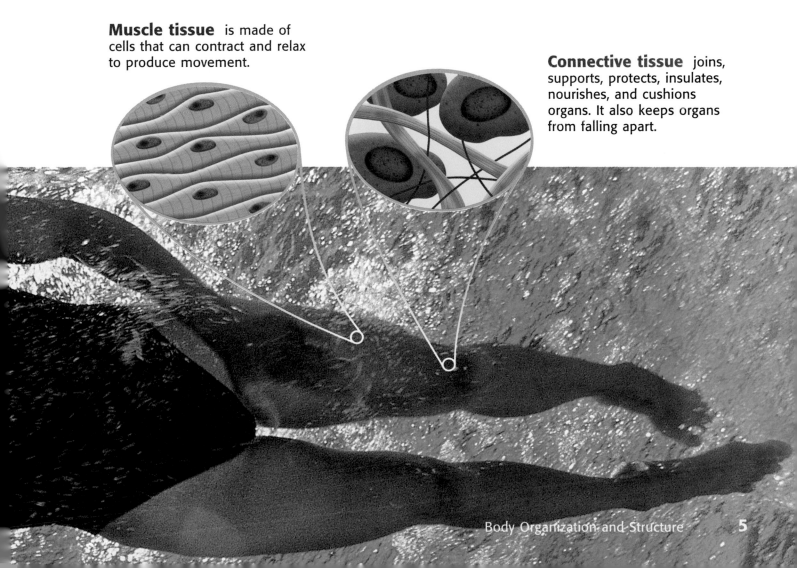

Muscle tissue is made of cells that can contract and relax to produce movement.

Connective tissue joins, supports, protects, insulates, nourishes, and cushions organs. It also keeps organs from falling apart.

Organic Systems

Organ Systems

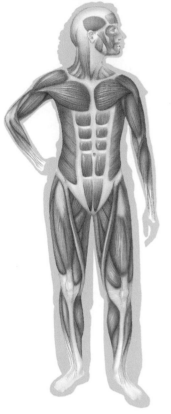

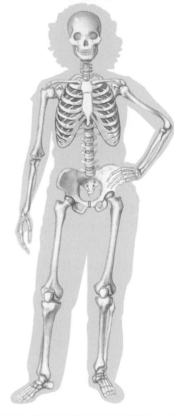

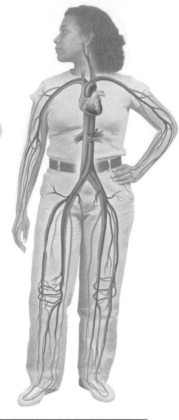

Integumentary system

Your skin, hair, and nails protect underlying tissue.

Muscular system

Your skeletal muscles move your bones.

Skeletal system

Your bones provide a frame to support and protect body parts.

Cardiovascular system

Your heart pumps blood through all your blood vessels.

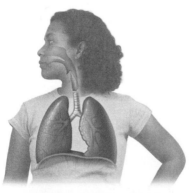

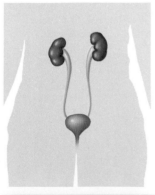

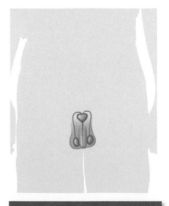

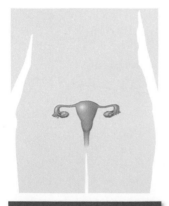

Respiratory system

Your lungs absorb oxygen and release carbon dioxide.

Urinary system

Your urinary system removes wastes from the blood and regulates the body's fluids.

Reproductive system (male)

The male reproductive system produces and delivers sperm.

Reproductive system (female)

The female reproductive system produces eggs and nourishes and shelters the unborn baby.

Organ Systems

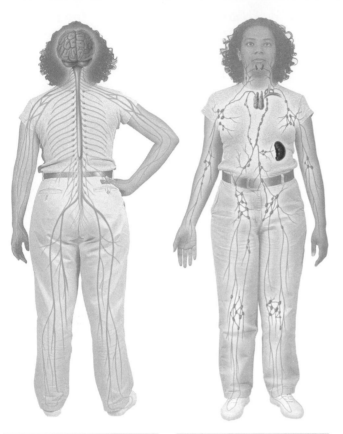
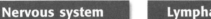

Nervous system

It is the role of the nervous system to receive and send electrical messages throughout the body.

Lymphatic system

Your lymphatic system returns leaked fluids to blood vessels. It also helps you get rid of germs that can harm you.

Digestive system

Your digestive system breaks down the food you eat into nutrients that can be absorbed into your body.

Endocrine system

Glands regulate body functions by sending out chemical messengers. The ovaries, in females, and testes, in males, are part of this system.

SECTION REVIEW

1. Explain the relationship between cells, tissues, organs, and organ systems.

2. Compare the four kinds of tissue found in the human body.

3. **Using Graphics** Make a chart that lists the major organ systems and their functions.

4. **Relating Concepts** Describe a time when homeostasis was disrupted in your body. Which body systems do you think were affected?

---

internet**connect**

SCI**LINKS**
NSTA

TOPIC: Tissues and Organs, Body Systems
GO TO: www.scilinks.org
sciLINKS NUMBER: HSTL530, HSTL535

SECTION 2
READING WARM-UP

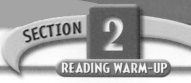

The Skeletal System

Terms to Learn

skeletal system
compact bone
spongy bone
cartilage
joint
ligament

What You'll Do

- Identify the major organs of the skeletal system.
- Describe the functions of bones.
- Illustrate the internal structure of bones.
- Compare three types of joints.
- Discuss how bones function as levers.

When you hear the word *skeleton,* you may think of the remains of something that has died. But your skeleton is not dead; it is very much alive. Your bones are not dry and brittle. They are just as alive and active as the muscles that are attached to them. Bones, cartilage, and the special structures that connect them make up your **skeletal system.**

The Burden of Being a Bone

Bones do a lot more than just hold you up. Your bones perform several important functions inside your body. The names of some of your bones are identified in **Figure 3.**

Protection Your heart and lungs are shielded by your ribs, your spinal cord is protected by your vertebrae, and your brain is protected by your skull.

Storage Bones store minerals that help the nerves and muscles function properly. Your arm and leg bones also store fat that can be used for energy.

Movement Skeletal muscles pull on the bones to produce movement. Without bones, you would not be able to sit, stand, walk, or run.

Blood Cell Formation Some of your bones are filled with a special material that makes blood cells.

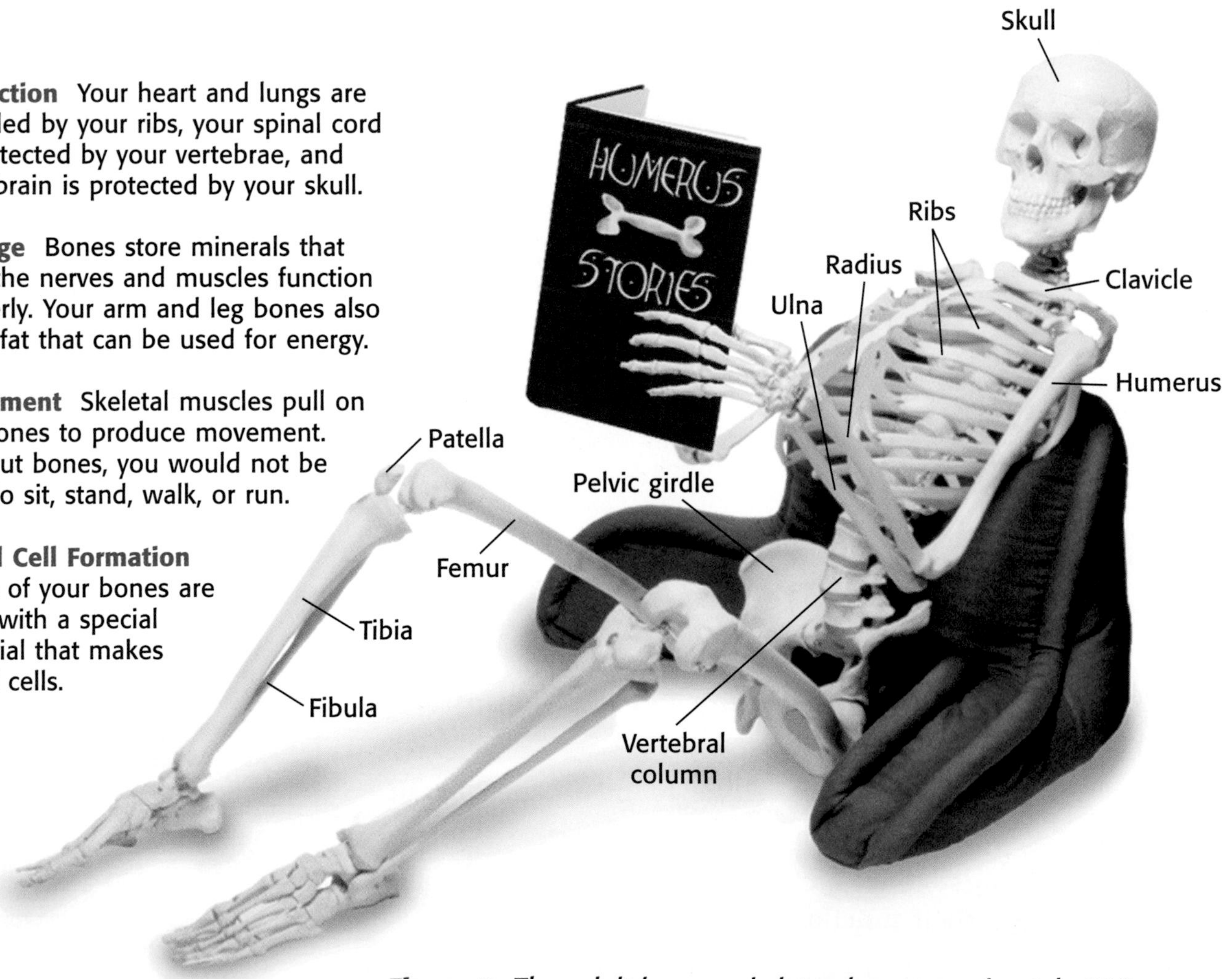

Skull
HUMERUS
STORIES
Ribs
Radius
Clavicle
Ulna
Humerus
Patella
Pelvic girdle
Femur
Tibia
Fibula
Vertebral column

Figure 3 *The adult human skeleton has approximately 206 bones. Several major bones are identified in this skeleton.*

What's in a Bone?

A bone may seem lifeless, but it is a living organ made of several different tissues. Bone is composed of connective tissue and minerals that are deposited by living cells called *osteoblasts*.

If you look inside a bone, you will notice there are two different kinds of bone tissue. If the tissue does not have any visible open spaces, it is called **compact bone.** Bone tissue that has many open spaces is called **spongy bone.** Spongy bone provides most of the strength and support for a bone. It acts like the trusses of a bridge.

Down to the Marrow Bones contain a soft tissue called *marrow*. Red marrow, sometimes found in spongy bone, produces red blood cells. Yellow marrow, found in the central cavity of long bones, stores fat. Tiny canals within the compact bone contain small blood vessels. **Figure 4** shows a cross section of a femur.

Pickled Bones

This activity lets you see how a bone changes when it is exposed to an acid, such as vinegar. Place a **clean chicken bone** in a **jar of vinegar.** After 1 week, remove the bone and rinse it with water. Make a list of changes that you can see or feel. How has the bone's strength changed? What did the vinegar remove?

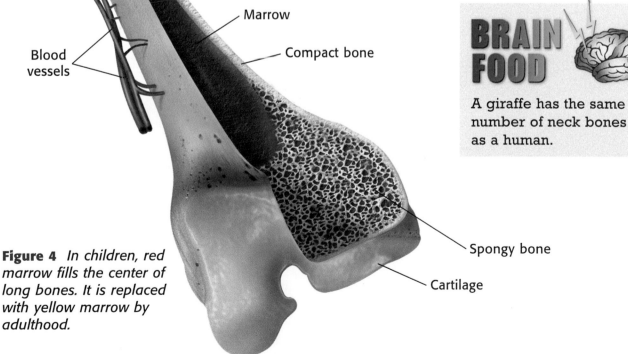

Marrow

Compact bone

Blood vessels

Spongy bone

Cartilage

Figure 4 *In children, red marrow fills the center of long bones. It is replaced with yellow marrow by adulthood.*

BRAIN FOOD

A giraffe has the same number of neck bones as a human.

Growing Bones

Did you know that most of your skeleton used to be soft and rubbery? Most bones start out as a soft, flexible tissue called **cartilage.** When you were born, you had little true bone. But as you grew, the cartilage was replaced by bone. During childhood, growth plates of cartilage remain in most bones, providing a place for those bones to continue to grow.

Feel the end of your nose, or bend the top of your ear. As shown in **Figure 5,** some areas, like these, never become bone. The flexible material beneath your skin in these areas is cartilage.

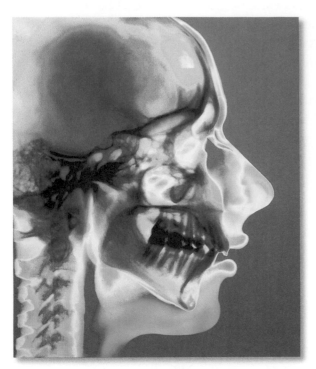

Figure 5 *The skull and neck bones in this computer-colored X ray are shown mostly in blue.*

What's the Point of a Joint?

The place where two or more bones connect is called a **joint.** Your joints have special designs that allow your body to move when your muscles contract. Some joints allow a lot of movement, while other joints are fixed, which means they allow little or no movement. For example, the joints in the skull are fixed. Joints that have a wide range of movement tend to be more susceptible to injury than those that are less flexible. Some examples of movable joints are shown in **Figure 6.**

Figure 6 *Joints are shaped according to their function in the body.*

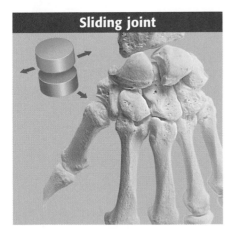

Sliding joint

Sliding joints allow bones in the hand to glide over one another, giving some flexibility to the area.

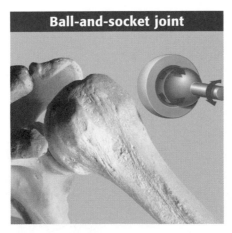

Ball-and-socket joint

Like a joystick on a computer game, the shoulder enables your arm to move freely in all directions.

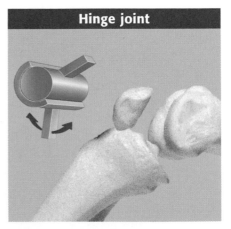

Hinge joint

Like a hinge on a door, the knee enables you to flex and extend your lower leg.

Bone to Bone Joints are kept together with strong elastic bands of connective tissue called **ligaments.** If a ligament is stretched too far, it becomes strained. A strained ligament will usually heal with time, but a torn ligament will not. A torn ligament must be repaired surgically. Cartilage helps cushion the area where two bones meet. If cartilage wears away, the joint becomes arthritic.

Can Levers Lessen Your Load?

You may not think of your limbs as being machines, but they are. The action of a muscle pulling on a bone often works like a type of simple machine called a *lever.* A lever is a rigid bar that moves on a fixed point known as a *fulcrum.* Any force applied to the lever is called the *effort.* A force that resists the motion of the lever, such as the downward force exerted by a weight on the bar, is called the *load* or the *resistance.* **Figure 7** shows how three types of levers are used in the human body.

Figure 7 *There are three classes of levers, based on the location of the fulcrum, the load, and the effort.*

First-class lever

The fulcrum lies between the load and the effort.

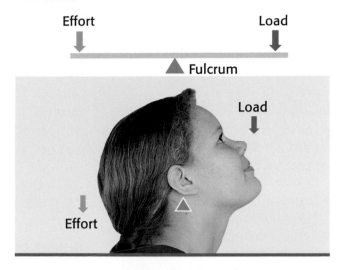

Second-class lever

The load lies between the fulcrum and the effort.

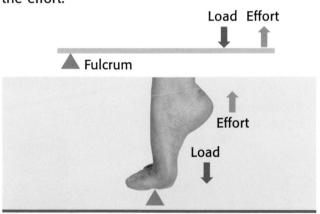

Third-class lever

The effort lies between the fulcrum and the load.

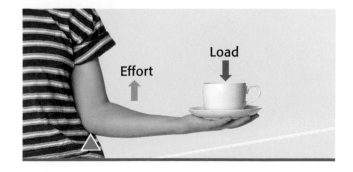

SECTION REVIEW

1. Describe four important functions of bones.

2. Draw a bone, and label the inside and outside structures. Use colored pencils to color and label spongy bone, blood vessels, marrow cavity, compact bone, and cartilage.

3. List three hinge joints in your body.

4. **Interpreting Models** Study the models of levers pictured in Figure 7. Use a small box (load), a ruler (bar), and a pencil (fulcrum) to create models of each type of lever.

The Muscular System

Terms to Learn

muscular system   skeletal muscle
smooth muscle   tendon
cardiac muscle

What You'll Do

◆ List the major parts of the muscular system.
◆ Describe the different types of muscle.
◆ Describe how skeletal muscles move bones.
◆ Compare aerobic exercise with resistance exercise.
◆ Give an example of a muscle injury.

Have you ever tried to be perfectly still for just 1 minute? Try as you might, you just can't do it. Somewhere in your body, certain muscles are always working. For example, muscles continuously push blood through your blood vessels. A muscle makes you breathe. And muscles hold you upright. If all your muscles rested at the same time, you would collapse. Your muscles are made of muscle tissue and connective tissue. Muscles that attach to bones and the connective tissue that attaches them make up the **muscular system.**

Types of Muscle

There are three types of muscle tissue that make up the muscles in your body. **Smooth muscle** is found in the digestive tract and the blood vessels. **Cardiac muscle** is a special type of muscle found only in your heart. **Skeletal muscles** are attached to your bones for movement, and they help protect your inner organs. The three types of muscles are shown in **Figure 8.**

    Muscle action can be voluntary or involuntary. Muscle action that is under your control is *voluntary.* Muscle action that is not under your control is *involuntary.* The actions of smooth muscle and cardiac muscle are involuntary. The actions of skeletal muscles can be both voluntary and involuntary. For example, you can blink your eyes any time you want to, but your eyes will also blink automatically if you do not think about it.

Figure 8 *Your body has smooth muscle, cardiac muscle, and skeletal muscle.*

Skeletal muscle enables bones to move.

Smooth muscle moves food through the digestive system.

Cardiac muscle causes the heart to beat.

Making Your Move

Skeletal muscles produce hundreds of different voluntary movements. This is demonstrated by a ballet dancer, a swimmer, or even someone making a funny face, as shown in **Figure 9.** When you want to make a movement, you cause electrical signals to travel from the brain to the skeletal muscle cells. The muscle cells respond to these signals by contracting or getting shorter.

Figure 9 *It takes an average of 13 muscles to smile and an average of 43 muscles to frown.*

Muscles to Bones Strands of tough connective tissue called **tendons** connect your skeletal muscles to your bones. When a muscle gets shorter, a pulling action occurs, bringing the bones closer to each other. For example, the biceps muscle, shown in **Figure 10,** is attached by tendons to a bone in your shoulder and to another bone in your forearm. When the biceps contracts, your arm bends.

Working in Pairs Your skeletal muscles work in pairs to cause smooth, controlled movements. Many basic movements are the result of muscle pairs that cause bending and straightening. If a muscle bends part of your body, then that muscle is called a *flexor.* If the muscle straightens part of your body, then it is called an *extensor.* The flexor muscle of the arm is the biceps. The extensor muscle of the arm is the triceps. Discover some of your own flexor and extensor muscles by doing the QuickLab at right.

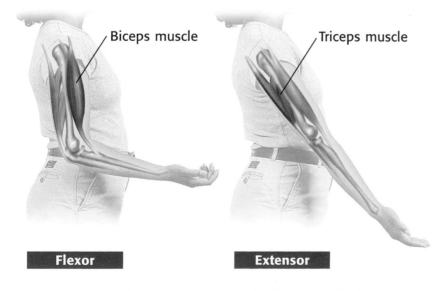

Biceps muscle · Triceps muscle

Flexor · **Extensor**

Figure 10 *Skeletal muscles, such as the biceps and triceps muscles, work in pairs. When the biceps muscle contracts, the elbow bends. When the triceps muscle contracts, the elbow straightens.*

QuickLab

Power in Pairs

1. While sitting in a chair, place one of your hands palm up under the edge of a **table.** Apply gentle upward pressure.
2. With your free hand, feel the front and back of your upper arm.
3. Next place your hand palm down on top of the table. Apply pressure downward.
4. Again with your free hand, feel the front and back of your upper arm.
5. What did you notice when you were pressing up? when you were pressing down?

TRY at HOME

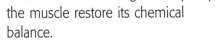

Chemistry
C O N N E C T I O N

Body chemistry is very important for healthy muscle functioning. If there is a chemical imbalance in a muscle due to excessive sweating, poor diet, tension, or illness, spasms or cramping may occur. Sodium, calcium, and potassium—three chemicals called *electrolytes*—must be in proper balance to avoid cramps and spasms. Relaxation and massage usually help the muscle restore its chemical balance.

Use It or Lose It

When someone breaks an arm and has to wear a cast, the muscles surrounding the injured bone change. That's because these muscles are not exercised, and they become smaller and weaker. On the other hand, exercised muscles are stronger and larger. Certain exercises can give muscles more endurance. This means they're able to work longer without getting tired. Strong muscles benefit other systems in your body too. When a muscle contracts, blood vessels in that muscle get squeezed. This helps push blood along, increasing blood flow without demanding more work from the heart.

Resistance Exercises To develop the size and strength of your skeletal muscles, resistance exercises are the most effective form of exercise. Resistance exercises require muscles to overcome the resistance (weight) of another object. Some resistance exercises, like the bent knee curl-up shown in **Figure 11,** require you to overcome your own weight.

Figure 11 *Resistance exercises are tough, but they can really help you build strong muscles.*

Aerobic Exercise Steady, moderate-intensity activity, such as jogging, cycling, skating, swimming laps, or walking, is called aerobic exercise. Aerobic exercise increases the size and strength of your skeletal muscles somewhat, but mostly it strengthens the heart while increasing the endurance of your skeletal muscles. Many people, like the girl in **Figure 12,** enjoy doing aerobic exercise.

Figure 12 *Aerobic exercise is a great way to have fun while strengthening your heart.*

✓ Self-Check

Which kind of skeletal muscle do you use to perform a curl-up? Which kind do you use to do a push-up? *(See page 212 to check your answers.)*

Muscle Injury

Any exercise program should be started gradually so that the muscles gain strength and endurance without injury. Muscles should also be warmed up gradually to reduce the risk of injury. However, as shown in **Figure 13,** the muscular system can experience damage. A muscle strain, commonly called a pulled muscle, is the overstretching or even tearing of a muscle. Muscle strain often occurs because the muscle has not been properly conditioned for the work it is doing.

Tendons, as well as muscles, can get injured from overuse. A damaged tendon can become hot or inflamed as your body tries to repair it. This painful condition is called tendinitis, and an extended period of rest is often required for the tendon to heal.

Figure 13 *A pulled hamstring is a tear or strain of one of the muscles or tendons on the back of the thigh.*

The Dangers of Anabolic Steroids Some people try to make their muscles larger and stronger by taking hormones called *anabolic steroids.* Anabolic steroids are powerful chemicals that resemble testosterone, a male sex hormone. Using anabolic steroids not only gives athletes an unfair advantage in competition but also puts the user at risk for serious long-term health problems. The use of anabolic steroids threatens the heart, liver, and kidneys, and it can cause high blood pressure. If taken before the skeleton is mature, anabolic steroids can cause the bones to stop growing.

MATH BREAK

Runner's Time

Jan, who has been a runner for several years, has decided to enter a race. She now runs 5 km in 30 minutes. She would like to decrease her time by 15 percent before the race. What will her time be when she meets her goal?

SECTION REVIEW

1. List three types of muscle tissue, and describe their functions in the body.

2. Compare aerobic exercise with resistance exercise, and give two examples of each.

3. **Applying Concepts** Describe the muscle action required to pick up a book. Make a sketch that illustrates the muscle action.

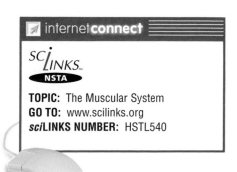

internet**connect**

SC*i*LINKS.
NSTA

TOPIC: The Muscular System
GO TO: www.scilinks.org
*sci*LINKS NUMBER: HSTL540

Terms to Learn

integumentary   epidermis
  system   dermis
sweat glands   hair follicle
melanin

What You'll Do

◆ Describe the major functions of the integumentary system.
◆ List the major parts of the skin, and discuss their functions.
◆ Describe the structure and function of hair and nails.
◆ Describe some common types of damage that can affect skin.

The Integumentary System

Here's a quiz for you. What part of your body has to be partly dead to keep you alive? Here are some clues: it comes in a variety of colors, it is the largest organ in the body, and it protects you from the outside world. Oh, and guess what—it is showing right now. Did you guess your skin? If you did, you guessed correctly.

Your skin, hair, and nails make up your **integumentary** (in TEG yoo MEN tuhr ee) **system.** (*Integument* means "covering.") Like all organ systems, the integumentary system helps your body maintain a healthy internal environment.

The Skin: More than Just a "Coat"

Why do you need skin? Here are four good reasons:

- Skin protects you by keeping moisture in your body and foreign particles out of your body.

- Skin keeps you "in touch" with the outside world. The nerve endings in your skin allow you to feel what's around you.

- Skin helps regulate your body's temperature. For example, small organs in the skin called **sweat glands** produce sweat, a salty liquid that flows to the surface of the skin. As sweat evaporates, the skin cools.

- Skin helps get rid of wastes. Several types of waste chemicals can leave the bloodstream and be removed in sweat.

What Determines Skin Color? A darkening chemical in skin called **melanin** determines skin color, as shown in **Figure 14.** If a lot of melanin is present, the skin is very dark. If only a little melanin is produced, the skin is very light. Melanin in the upper layer of the skin absorbs much of the harmful radiation from the sun, reducing DNA damage that can lead to cancer. However, *all* skin is vulnerable to cancer and therefore should be protected from sun exposure whenever possible.

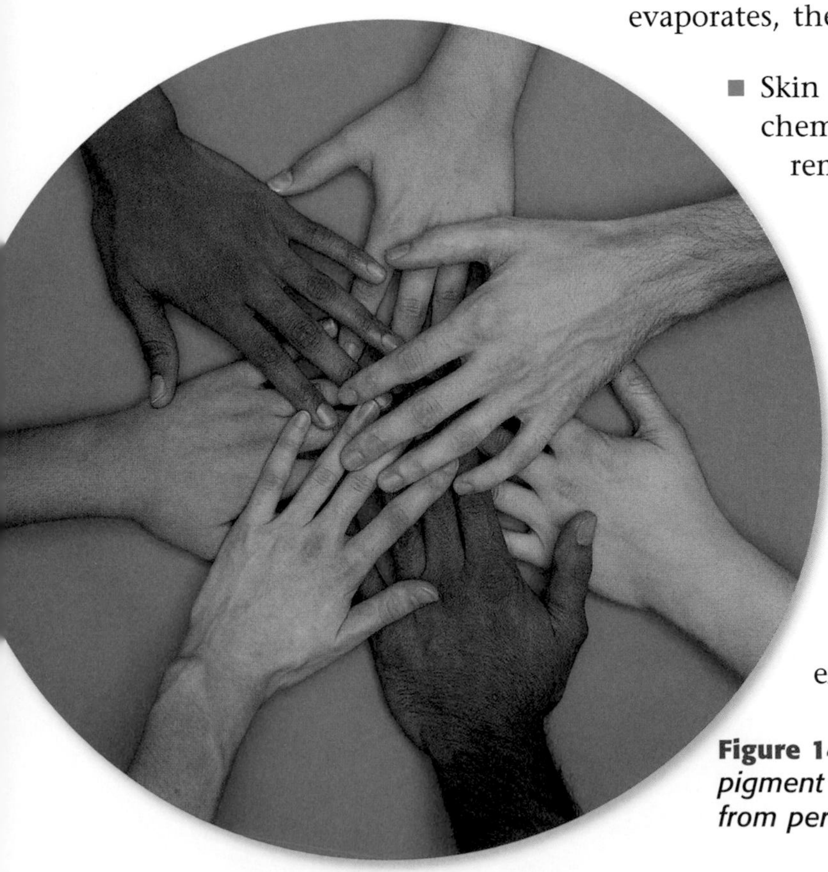

Figure 14 *Variety in skin color is caused by the pigment melanin. The amount of melanin varies from person to person.*

A Tale of Two Layers

As you already know, the skin is the largest organ of your body. In fact, the skin of an adult covers an area of about 2 m²! However, there's a lot more to skin than meets the eye. The skin has two main layers: the dermis and the epidermis. The **epidermis** is the thinner layer of the two. It's what you see when you look at your skin. (*Epi* means "on top of.") The deeper, thicker layer is known as the **dermis.**

Epidermis The epidermis is composed of a type of epithelial tissue. Even though the epidermis has many layers of cells, it is only as thick as two sheets of notebook paper over most of the body. It is thicker in the palms of your hands and the soles of your feet. Most epidermal cells are dead and are filled with a protein called keratin, which helps make the skin tough.

Dermis The dermis lies underneath the epidermis. It is mostly connective tissue, and it contains many fibers made of a protein called collagen. The fibers provide strength and allow skin to bend without tearing. The dermis also contains a variety of small structures, as shown in **Figure 15.**

✔ Self-Check

To what system do the skin's blood vessels belong? (See page 212 to check your answer.)

Your epidermis is showing!

Figure 15 *Beneath the surface, your skin is a complex organ made of blood vessels, nerves, glands, and muscles.*

Blood vessels transport substances and help regulate body temperature.

Nerves carry messages to and from the brain.

Muscle fibers attached to a hair follicle can contract, causing the hair to stand up.

Hair follicles in the dermis produce hair.

Oil glands release oil that keeps hair flexible and helps waterproof the epidermis.

Sweat glands release sweat. As sweat evaporates, heat is removed from the skin, and the body is cooled. Sweat also contains waste materials taken out of the body.

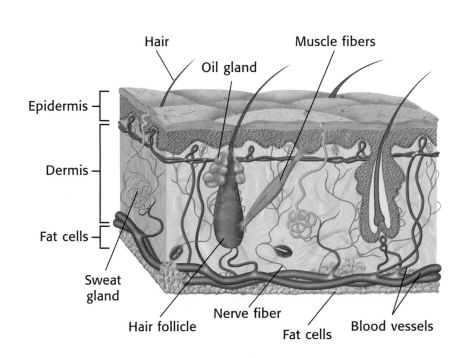

How fast do your fingernails grow? Find out on *page 178.*

Hair and Nails

A hair, shown in **Figure 16,** is formed at the bottom of a tiny sac called a **hair follicle.** The hair grows as new cells are added at the hair follicle and older cells get pushed upward. The only living cells in a hair are in the hair follicle, where the hair is produced.

Letting Your Hair Down Hairs protect skin from ultraviolet light and can help keep particles, such as dust and insects, out of your eyes and nose. Like skin, hair gets its color from the pigment melanin. Dark hair contains more melanin than blond hair. In most mammals, hair also helps regulate body temperature. A contraction of a tiny muscle attached to the hair follicle causes the follicle to bend. In humans, the bending follicle pushes up the epidermis to make a goose bump. If the follicle contains a hair, the hair "stands up." The lifted hairs function like a sweater to trap warm air around the body.

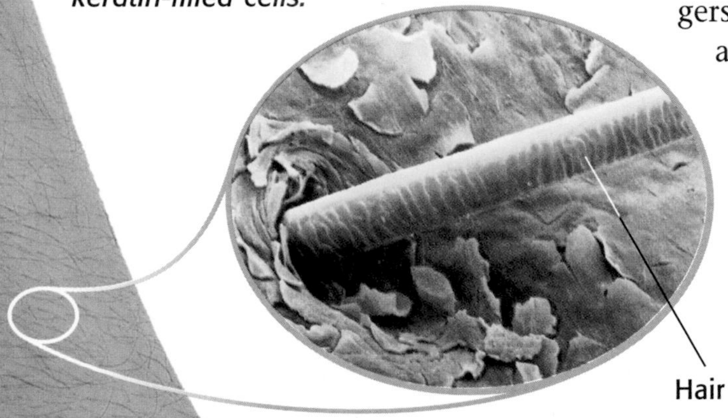

Figure 16 *A hair is actually layers of dead, tightly packed, keratin-filled cells.*

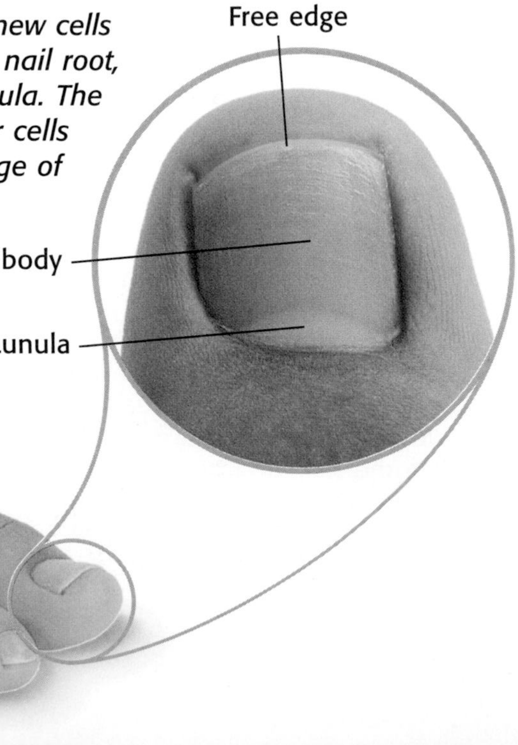

A Nail Tale Nails protect the tips of your fingers and toes so that they can remain soft and sensitive. This allows you to have a keen sense of touch. Nails form from *nail roots* under the skin at the base and sides of nails. As new cells form, the nail grows longer. The parts of a nail are shown in **Figure 17.**

Hair

Figure 17 *In nails, new cells are produced in the nail root, just beneath the lunula. The new cells push older cells toward the outer edge of the nail.*

Free edge

Nail body

Lunula

Living in Harm's Way

Skin is often damaged. The damage may be minor—a blister, an insect bite, or a small cut. Fortunately, your skin has an amazing ability to repair itself, as shown in **Figure 18.**

Figure 18 How Skin Heals

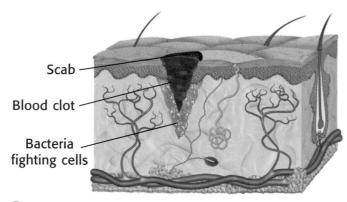

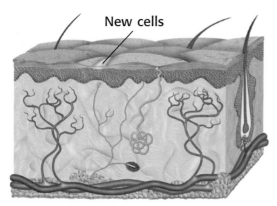

1 When you get a cut, a blood clot forms to prevent bacteria from entering the wound. Bacteria-fighting cells then come to the area to kill bacteria.

2 Damaged cells are replaced through cell division. Eventually, all that is left on the surface is a scar.

Other damage to the skin is very serious. Damage to the genetic material in skin cells can result in uncontrolled cell division, producing a mass of skin cells called a tumor. The term *cancer* is used to describe a tumor that invades other tissue. Darkened areas on the skin, such as moles, should be watched carefully for signs of cancer. **Figure 19** shows an example of a mole that has possibly become cancerous.

Your skin may also be affected by hormones that cause the oil glands in your skin to produce excess oil. This oil combines with dead skin cells and bacteria to clog hair follicles and cause infections. Proper cleansing and daily skin care can be helpful in decreasing the amount of infections.

Figure 19 *This mole has two halves that do not match, a characteristic that might indicate skin cancer.*

SECTION REVIEW

1. Why does skin color vary from person to person?

2. List six structures found in the dermis and the function of each one.

3. **Making Inferences** Why do you feel pain when you pull on your hair or nails but not when you cut them?

internet**connect**

SC*L*INKS.
NSTA

TOPIC: Integumentary System
GO TO: www.scilinks.org
*sci*LINKS NUMBER: HSTL545

Design Your Own Lab

Muscles at Work

Have you ever exercised outside on a cold fall day, wearing only a thin warm-up suit or shorts? How did you stay warm? The answer is that your muscle cells contracted. When contraction takes place, some energy is used to do work and the rest is converted to thermal energy. The thermal energy helps your body maintain a constant temperature in cold conditions. When you exercise strenuously on a hot summer day, your muscles can cause your body to become overheated. In this activity, your job is to find out how the release of energy can cause a change in your body temperature.

MATERIALS

- clock or watch with a second hand
- small hand-held thermometer
- other materials as approved by your teacher

Ask a Question

1 Form a group of four students. In your group, discuss what you already know about how muscle contractions can affect body temperature. During your discussion, ask several questions about how the release of energy can cause a change in body temperature. As a group, pick one of the questions that you think you can answer by performing an experiment.

Form a Hypothesis

2 Formulate a testable hypothesis to answer the question you chose. Write your hypothesis in your ScienceLog.

3 Plan an investigative procedure that includes the steps that are necessary to test your hypothesis. You will need to select the appropriate equipment to use during your experiment. Be sure to get your teacher's approval before you begin.

Conduct an Experiment

4 Assign tasks such as note taking, data recording, and timing to individuals in the group. What observations and data will you be recording? Design tables using a computer or graph paper to organize and examine the data you collect.

5 Perform your experiment as planned by your group. Be sure to record in your data tables all of the observations made during the experiment.

Analyze the Results

6 After you complete your experiment, review the data that you collected. Use a computer or graph paper to organize the data into graphs and charts.

7 Using your tables, charts, and graphs, can you make any inferences about how muscle contractions affect body temperature?

8 Do you recognize any patterns in your data? What trends can you predict? What might happen to your body temperature when you sleep?

Draw Conclusions

9 Was your hypothesis supported by your data? Communicate your conclusions in a written report. Describe how you could improve your experimental method.

Going Further
Why do humans shiver in the cold? Do all animals shiver? Find out why shivering is one of the first signs that your body is becoming too cold.

Chapter Highlights

Vocabulary

homeostasis *(p. 4)*

tissue *(p. 4)*

epithelial tissue *(p. 4)*

nervous tissue *(p. 4)*

muscle tissue *(p. 5)*

connective tissue *(p. 5)*

organ *(p. 5)*

Section Notes

- Your body maintains a stable internal environment called homeostasis.

- Four types of tissues work to maintain homeostasis. Each tissue has a special job to do.

- Tissues work together to form organs.

- A group of organs working together for a common purpose is called an organ system.

- There are 11 major organ systems in the human body.

Vocabulary

skeletal system *(p. 8)*

compact bone *(p. 9)*

spongy bone *(p. 9)*

cartilage *(p. 10)*

joint *(p. 10)*

ligament *(p. 11)*

Section Notes

- The skeletal system includes bones, cartilage, and ligaments.

- Bones support and protect the body, store minerals and fat, and produce blood cells.

- A typical bone contains marrow, spongy bone, compact bone, blood vessels, and cartilage.

☑ Skills Check

Math Concepts

CALCULATING A PERCENTAGE In the MathBreak on page 15 you were asked to calculate a percentage of a number. To do this, first express the percentage as a decimal or a fraction. Then multiply it by the number. For example, 25 percent can be written as 0.25 or 25÷100. To find 25 percent of 48, multiply by either 0.25 or 25÷100.

$$0.25 \times 48 = 12$$
or
$$(25 \div 100) \times 48 = 12$$

Visual Understanding

MOVING WITH JOINTS Take another look at the three kinds of joints on page 10. Consider how your joints work when you throw a ball or walk up stairs. The hinge joint in your knee can move freely in only two directions. The ball-and-socket joint in your shoulder can move in many directions. The sliding joints in your hand allow bones to glide over one another.

SECTION 2

- A joint is where two bones meet. Some joints allow a lot of movement, and some allow little or no movement.
- Bones are attached to bones by connective tissue called ligaments.
- The action of muscle on bone and joints often works like a simple machine called a lever.

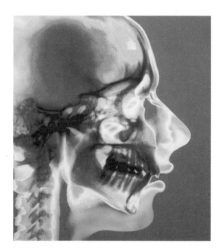

SECTION 3

Vocabulary

muscular system *(p. 12)*
smooth muscle *(p. 12)*
cardiac muscle *(p. 12)*
skeletal muscle *(p. 12)*
tendon *(p. 13)*

Section Notes

- Skeletal muscles and tendons make up the muscular system.
- You have three types of muscle: smooth, cardiac, and skeletal.
- Muscles are attached to bones by tendons.
- Exercise helps keep your muscular system healthy.

SECTION 4

Vocabulary

integumentary system *(p. 16)*
sweat glands *(p. 16)*
melanin *(p. 16)*
epidermis *(p. 17)*
dermis *(p. 17)*
hair follicle *(p. 18)*

Section Notes

- Your skin, hair, and nails make up your integumentary system.
- Your skin has two layers that contain a variety of small organs.
- Your hair and nails help protect your body.
- Skin can be damaged, but it has an amazing ability to repair itself.

Labs

Seeing Is Believing *(p. 178)*

---

internetconnect

GO TO: go.hrw.com

Visit the **HRW** Web site for a variety of learning tools related to this chapter. Just type in the keyword:

KEYWORD: HSTBD1

GO TO: www.scilinks.org

Visit the **National Science Teachers Association** on-line Web site for Internet resources related to this chapter. Just type in the *sci*LINKS number for more information about the topic:

TOPIC: Tissues and Organs   *sci*LINKS NUMBER: HSTL530
TOPIC: Body Systems   *sci*LINKS NUMBER: HSTL535
TOPIC: Skeletal System   *sci*LINKS NUMBER: HSTL537
TOPIC: The Muscular System   *sci*LINKS NUMBER: HSTL540
TOPIC: Integumentary System   *sci*LINKS NUMBER: HSTL545

Chapter Review

To complete the following sentences, choose the correct term from each pair of terms listed below:

1. Electrical signals are sent throughout the body by the __?__ tissue. (*epithelial* or *nervous*)

2. Your __?__ system is made up of skin, hair, and nails. (*integumentary* or *muscular*)

3. Bones are moved by __?__ muscle. (*smooth* or *skeletal*)

4. When __?__ muscles contract, they cause parts of the body to bend. (*extensor* or *flexor*)

5. Most of the skeleton starts out as __?__, which is later replaced by bone. (*cartilage* or *ligaments*)

UNDERSTANDING CONCEPTS

Multiple Choice

6. Which of the following is made up of cells that can contract and relax?
   a. skeletal tissue
   b. muscle tissue
   c. connective tissue
   d. nervous tissue

7. The organ system that provides support and protection for body parts is the
   a. endocrine system.
   b. circulatory system.
   c. skeletal system.
   d. respiratory system.

8. The epidermis is composed of
   a. dermis.
   b. epithelial tissue.
   c. connective tissue.
   d. true skin.

9. The fixed point in a lever is the
   a. effort.
   b. load.
   c. fulcrum.
   d. mechanical advantage.

10. Muscles cause bones to move when
    a. the muscles stretch.
    b. the muscles grow between bones.
    c. the muscles pull on bones.
    d. the muscles push bones apart.

11. Ligaments are the connective tissue that attaches
    a. bones to muscles.
    b. bones to other bones.
    c. muscles to other muscles.
    d. muscles to dermis.

Short Answer

12. Summarize the functions of the four types of tissues, and draw a sketch of each type.

13. How does the skin help protect the body?

14. What is a goose bump?

15. What are two ways skeletal muscle differs from cardiac muscle?

16. How do the functions of the skeletal system relate to the functions of the muscular system?

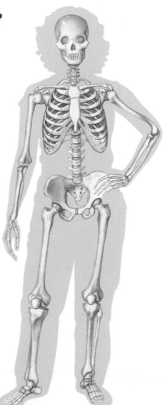

Concept Mapping

17. Use the following terms to create a concept map: bones, marrow, skeletal system, spongy bone, compact bone, cartilage.

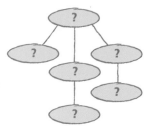

CRITICAL THINKING AND PROBLEM SOLVING

Write one or two sentences to answer the following questions:

18. Why do some muscles not work when a bone is broken?

19. Unlike human bones, some bird bones have air-filled cavities. What advantage does this give birds?

20. Compare the shapes of the bones of the human skull with the shapes of the bones of the human leg. Why is their shape important?

21. Compare the texture and sensitivity of the skin on your elbows with those of the skin on your fingertips. How can you explain the differences?

MATH IN SCIENCE

22. Your muscles make up about 40 percent of your overall mass. What is the muscle mass of a person whose total body mass is 60 kg?

23. The average person blinks 700 times an hour. How many times would the average person blink in a week if he or she were awake for 16 hours each day?

INTERPRETING GRAPHICS

Look at the picture below, and answer the questions that follow.

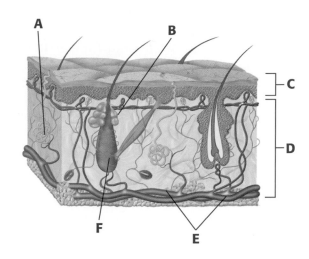

24. What is *D* called? What type of tissue is most abundant in this layer?

25. What is the name and function of *A*?

26. What is the name and function of *B*?

27. What part of the skin is made up of epithelial tissue that contains dead cells?

28. How does skin help regulate body temperature?

Reading Check-up

Take a minute to review your answers to the Pre-Reading Questions found at the bottom of page 2. Have your answers changed? If necessary, revise your answers based on what you have learned since you began this chapter.

Science, Technology, and Society

Engineered Skin

Your skin is more than just a well-fitting suit—it's your first line of defense against the outside world. Your skin keeps you safe from dehydration and infection, and the oil glands in your skin keep you waterproof. But what happens when a significant portion of skin is damaged?

More Skin Is the Answer

Sometimes doctors perform a skin graft, transferring some of a person's healthy skin to a damaged area of skin. This is because skin is really the best "bandage" for a wound. It protects the wound but still allows it to breathe. And unlike manufactured cloth or plastic bandages, skin can regenerate itself as it covers a wound. Sometimes, though, a person's skin is so severely damaged (as often occurs in burn victims) that the person doesn't have enough skin to spare.

Tissue Engineering

In the past few years, scientists have been studying tissue engineering to learn more about how the human body heals itself naturally. Using a small piece of young, healthy human skin and some collagen from cows, scientists can now engineer human skin. During the engineering process, cells form the dermal and epidermal layers of skin just as they would if they were still on the body. The living human skin that results can even heal itself if it is cut before it is used for a skin graft. Because it is living, the skin must be kept on a medium that provides it with nutrients until it is placed on a wound. Over time, the color of the grafted skin changes to match the color of the skin that surrounds it.

A Woven Dermis

Tissue engineers have also created another kind of skin, except this one has an unusual dermal

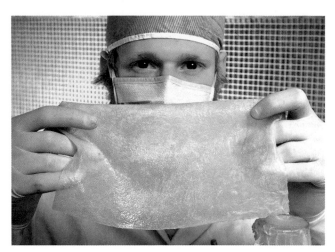

▲ *This is a piece of engineered skin used for grafting.*

and epidermal layer. In this skin, the dermis is made of woven collagen fibers. The wounded area digests these fibers and uses them as a guide to create a new dermis. The epidermal layer is a temporary layer of silicone. It shields the body from infection and protects against dehydration while new skin is being made.

After a new dermal layer forms under the protective silicone epidermis, the body is in better condition to accept a skin graft. Doctors can also graft a thinner portion of skin. A thinner graft is better for the body in the long run because it is easier to take from another part of the body. The new dermal layer also gives the body more time to strengthen on its own before the trauma of transplanting healthy skin to other areas.

On Your Own

▶ In the past, doctors have harvested skin from the bodies of people who, before they died, chose to be organ donors. What kinds of problems could arise if this harvested skin were used on burn victims?

Eureka!

Hairy Oil Spills

Oil and water don't mix, right? Oil floats on the surface of water and is clearly visible to the naked eye. Proving this in your kitchen isn't difficult, nor is it dangerous. But what happens when the water is the ocean and the oil is crude oil? You have an environmental disaster that costs millions of dollars to clean up. The worst example in American waters was in 1989 when the *Exxon Valdez* oil tanker spilled nearly 42 million liters of crude oil into the waters of Prince William Sound on the Alaskan coast.

▲ *Phil McCrory among bags of discarded human hair.*

▲ *This otter was drenched with oil spilled from the* Exxon Valdez.

Backyard Testing

A Huntsville, Alabama, hairdresser asked a brilliant question when he saw an otter whose fur was drenched with oil from the *Valdez* spill. If the otter's fur soaked up all the oil, why wouldn't human hair do the same? The hairdresser, Phil McCrory, gathered hair from the floor of his salon and took it home to perform his own experiments. He stuffed 2.2 kg of hair into a pair of his wife's pantyhose and tied the ankles together to form a bagel-shaped bundle. After filling his son's wading pool with water, McCrory floated the bundle in the pool. Next, McCrory poured used motor oil into the center of the ring. When he pulled the ring closed, not a drop of oil remained in the water!

How Does Hair Do This?

What McCrory discovered was that hair *adsorbs* oil instead of *absorbing* it. To adsorb means to collect a liquid or gas in layers on a surface. Because tiny cuticles cover every hair shaft like fish scales, the oil can bind to the surface of hair. Compare this process with the way a sponge works. A sponge completely absorbs a liquid. This means it is wet throughout, not just on the surface.

McCrory approached the National Aeronautics and Space Administration (NASA) with his discovery. In controlled tests performed by NASA, hair proved to be the fastest adsorber around. A little more than 1 kg of hair can adsorb over 3.5 L of oil in just 2 minutes!

It is estimated that within a week, 64 million kilograms of hair in reusable mesh pillows could have soaked up *all* of the oil spilled by the *Valdez*. Unfortunately, the $2 billion spent on the cleanup removed only about 12 percent of the spill. Did you ever think that the hair from your head could have a purpose beyond keeping your head warm?

Compare the Facts

▶ Research how McCrory's discovery compares with the methods currently used to clean up oil spills. Share your findings with the class.

Circulation and Respiration

Pre-Reading
Questions

1. What is blood, and what is its function in your body?
2. Why do you need to breathe?

SMALL BUT MIGHTY!

These donut-shaped objects are red blood cells like those that can be found throughout your body. Red blood cells are smaller than most other body cells. In fact, millions of them can be found in a single drop of blood. These cells may be small, but they perform a very important function. They are so important that your body makes about 200 billion new red blood cells every day. Why does your body need so many red blood cells? In this chapter, you will learn how these tiny cells enable all your body cells to carry out cellular respiration.

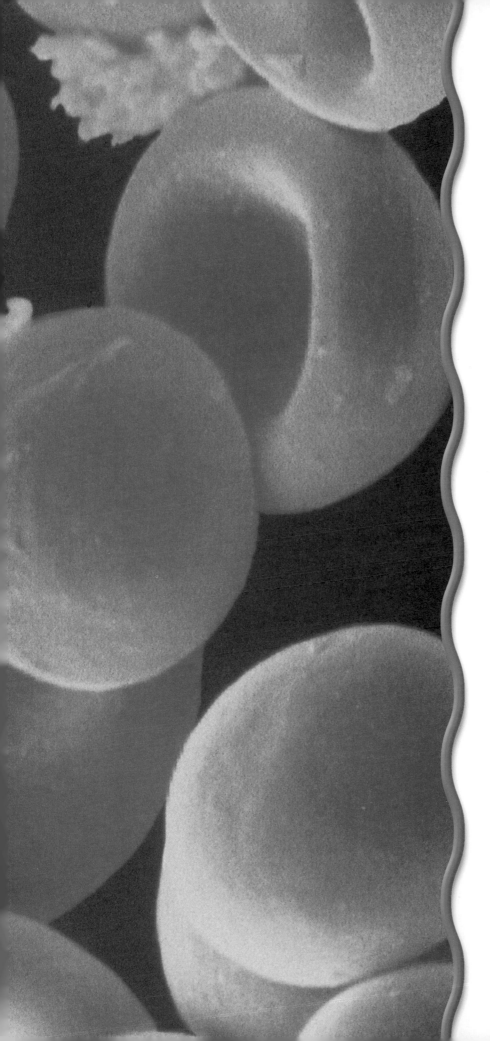

EXERCISE AND YOUR HEART

Your heart pumps blood throughout your body. How does your heart respond to exercise? You can determine this reaction by measuring your pulse. You can take your pulse by placing your fingers on the inside of your wrist just below your thumb.

Procedure

1. Take your pulse while remaining still. Using a **watch with a second hand,** count the number of beats in 15 seconds. Then multiply this number by 4 to calculate the number of beats in 1 minute

2. Do jumping jacks or jog in place for 30 seconds. Then stop and calculate your heart rate again.

    Caution: Do not perform this exercise if you have difficulty breathing, have high blood pressure, or easily get dizzy.

3. Rest for 5 minutes, and then take your pulse again.

Analysis

4. How did exercise affect your heart rate? Why do you think this happened?

5. How does your heart rate affect the rate at which red blood cells travel throughout your body?

6. Why did your heart rate return to normal after you rested?

Terms to Learn

cardiovascular system
blood
arteries
capillaries
veins
pulmonary circulation
systemic circulation
blood pressure

What You'll Do

◆ Describe the functions of the cardiovascular system.
◆ Compare and contrast the three types of blood vessels.
◆ Describe the path that blood travels as it circulates through the body.
◆ Distinguish between blood types.

The Cardiovascular System

When you hear the word *heart,* what do you think of first? Many people think of romance. But the heart is much more than a symbol of love. It's the pump that drives your cardiovascular system. The **cardiovascular system** transports materials to and from your cells. The word *cardio* means "heart," and the word *vascular* means "vessel." The cardiovascular system, which is shown in **Figure 1,** is made up of three parts: blood, the heart, and blood vessels.

What Is Blood?

The human body contains about 5 L of blood. **Blood** is a connective tissue made up of two types of cells, cell parts, and plasma. *Plasma* is the fluid part of blood. It is a mixture of water, minerals, nutrients, sugars, proteins, and other substances. Red blood cells, white blood cells, and platelets float in the plasma.

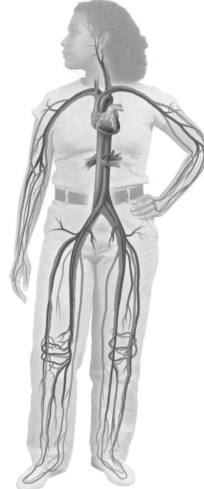

**Figure 1
The Cardiovascular System**

Figure 2 *Red blood cells deliver oxygen.*

Red Blood Cells Red blood cells, or RBCs, are the most abundant cells in blood. RBCs, shown in **Figure 2,** supply your cells with oxygen. As you have learned, cells need oxygen to carry out cellular respiration. Each RBC contains a protein called *hemoglobin* (HEE moh GLOH bin). Hemoglobin, which gives RBCs their red color, clings to the oxygen you inhale. This allows RBCs to transport oxygen throughout the body. The shape of RBCs gives them a large amount of surface area for absorbing and releasing oxygen.

RBCs are made in the bone marrow. Before RBCs enter the bloodstream, they lose their nucleus and other organelles. Without a nucleus, which contains DNA, the RBCs cannot replace worn-out proteins. RBCs therefore can live only about 4 months.

White Blood Cells Sometimes *pathogens*—bacteria, viruses, and other microscopic particles that can make you sick—are able to enter your body. When they do, they often encounter your white blood cells, or WBCs. WBCs, shown in **Figure 3,** help you stay healthy by destroying pathogens and helping to clean wounds.

WBCs fight pathogens in several ways. Some squeeze out of vessels and move around in tissues, searching for pathogens. When they find a pathogen, they engulf it. Other WBCs release chemicals called *antibodies,* which help destroy pathogens. WBCs also keep you healthy by engulfing body cells that have died or been damaged. WBCs are made in bone marrow. Some of them mature in lymphatic organs, which will be discussed later.

Platelets Drifting among the blood cells are tiny particles called platelets. *Platelets* are pieces of larger cells found in bone marrow. These larger cells remain in the bone marrow, but they pinch off fragments of themselves, which enter the blood. Although platelets last for only 5 to 10 days, they are an important part of blood. When you cut or scrape your skin, you bleed because blood vessels have been opened. As soon as bleeding occurs, platelets begin to clump together in the damaged area and form a plug that helps reduce blood loss, as shown in **Figure 4.** Platelets also release a variety of chemicals that react with proteins in the plasma and cause tiny fibers to form. The fibers create a blood clot.

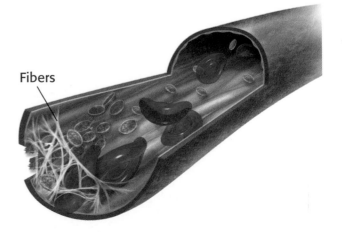

Figure 3 *White blood cells defend the body against pathogens. These white blood cells have been colored yellow to make their shape easier to see.*

Figure 4 *Platelets release chemicals in damaged vessels and cause fibers to form. The fibers make a "net" that traps blood cells and stops bleeding.*

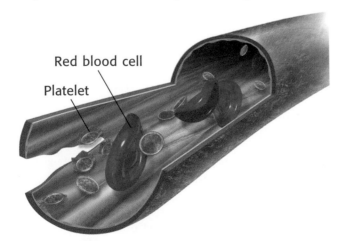

Red blood cell

Platelet

Fibers

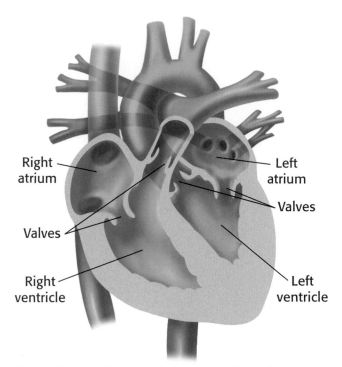

Right atrium

Left atrium

Valves

Valves

Right ventricle

Left ventricle

Figure 5 *The heart is a four-chambered organ that pumps blood through cardiovascular vessels. The vessels carrying oxygen-rich blood are shown in red. The vessels carrying oxygen-poor blood are shown in blue.*

Have a Heart

Your heart is a muscular organ about the size of your fist. It is found in the center of your chest cavity. The heart pumps oxygen-poor blood to the lungs and oxygen-rich blood to the body. Like the hearts of all mammals, your heart has a left side and a right side that are separated by a thick wall. As you can see in **Figure 5,** each side has an upper chamber and a lower chamber. Each upper chamber is called an *atrium* (plural, *atria*). Each lower chamber is called a *ventricle.*

Flaplike structures called *valves* are located between the atria and ventricles and also where large arteries are attached to the heart. As blood moves through the heart, the valves close and prevent blood from going backward. The lub-dub, lub-dub sound that a beating heart makes is caused by the closing of the valves. The flow of blood through the heart is shown in the diagram below.

The Flow of Blood Through the Heart

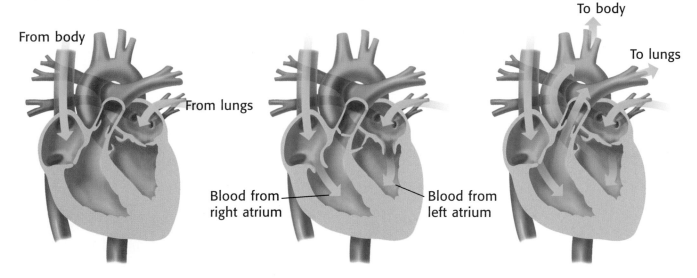

From body

From lungs

Blood from right atrium

To body

To lungs

Blood from left atrium

1 Blood enters the atria first. The left atrium receives oxygen-rich blood from the lungs. The right atrium receives oxygen-poor blood from the body.

2 When the atria contract, blood is squeezed into the ventricles.

3 While the atria relax, the ventricles contract and push blood out of the heart. Blood from the right ventricle goes to the lungs. Blood from the left ventricle goes to the rest of the body.

Blood Vessels

Blood travels throughout your body in blood vessels. A blood vessel is a hollow tube that transports blood. There are three types of blood vessels—arteries, capillaries, and veins. Their structures and their relationship to each other are shown in **Figure 6.**

Figure 6 *Large arteries branch into smaller arteries, which branch into capillaries. Capillaries join small veins, which join to form large veins.*

To heart

From heart

Vein

Capillaries

Artery

Arteries **Arteries** are blood vessels that direct blood away from the heart. Arteries have thick elastic walls that contain a layer of smooth muscle. Each time the heart beats, blood is pumped out of the heart at high pressure. The thick walls of arteries have the strength to withstand this pressure. The rhythmic change in blood pressure is called a *pulse.*

Capillaries A strand of hair is about 10 times wider than a capillary. **Capillaries** are the smallest blood vessels in your body. Capillary walls are only one cell thick. As shown in **Figure 7,** capillaries are so narrow that blood cells must pass through them in single file. The simple structure of a capillary allows nutrients, oxygen, and many other kinds of substances to diffuse easily through capillary walls. No cell in the body is more than three or four cells away from a capillary.

Veins After leaving capillaries, the blood enters veins. **Veins** are blood vessels that direct the blood back to the heart. As blood travels through veins, valves keep the blood from flowing backward. When skeletal muscles contract, they squeeze nearby veins and help push blood toward the heart.

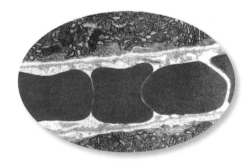

Figure 7 *These red blood cells are traveling through a capillary.*

BRAIN FOOD

If all the blood vessels in your body were strung together, the total length would be more than twice the circumference of the Earth.

Circulation and Respiration **33**

Going with the Flow

As you read earlier, one important function of your blood is to supply the cells of your body with oxygen. Where does blood get this oxygen? It gets it from your lungs during pulmonary circulation. **Pulmonary circulation** is the circulation of blood between your heart and lungs.

When oxygen-rich blood returns to the heart from the lungs, it must be pumped to the rest of the body. The circulation of blood between the heart and the rest of the body is called **systemic circulation.** Both types of circulation are diagramed below.

The Flow of Blood Through the Body

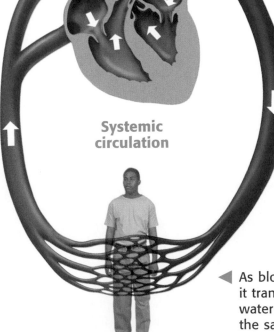

The right ventricle pumps ▶ oxygen-poor blood into arteries that lead to the lungs. These are the only arteries in the body that carry oxygen-poor blood.

Pulmonary circulation

◀ In the capillaries of the lungs, blood absorbs oxygen and releases carbon dioxide. Oxygen-rich blood travels through veins to the left atrium. These are the only veins in the body that carry oxygen-rich blood.

Oxygen-poor blood travels ▶ back to the heart and is delivered into the right atrium by two large veins.

Systemic circulation

◀ The heart pumps oxygen-rich blood from the left ventricle into arteries and then into capillaries.

◀ As blood travels through capillaries, it transports oxygen, nutrients, and water to the cells of the body. At the same time, waste materials and carbon dioxide are carried away.

Blood Flows Under Pressure

When you run water through a hose, you can feel the hose stiffen as the water pushes against the inside of the hose. Blood has the same effect on your blood vessels. The force exerted by blood on the inside walls of a blood vessel is called **blood pressure.**

Like the man shown in **Figure 8,** many people get their blood pressure checked on a regular basis. Blood pressure is reported in millimeters (mm) of mercury, Hg. A blood pressure of 120 mm Hg means the pressure on the vessel walls is great enough to push a narrow column of mercury 120 mm high.

A normal blood pressure is about 120/80. The first number is called the systolic pressure. *Systolic pressure* is the pressure inside large arteries when the ventricles contract. As you read earlier, the surge of blood causes the arteries to bulge and produce a pulse. The second number is called the diastolic pressure. *Diastolic pressure* is the pressure in the arteries when the ventricles relax.

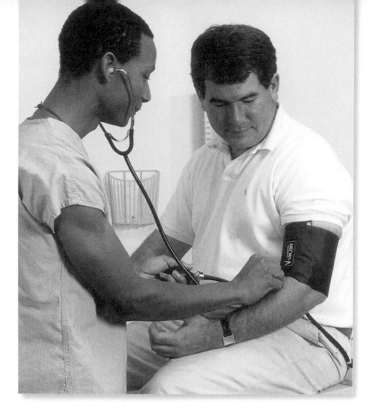

Figure 8 *This nurse is measuring a patient's blood pressure. Consistently high or low blood pressure may suggest a problem with the cardiovascular system.*

Exercise and Blood Flow

When you exercise, your muscles require much more oxygen and nutrients. To solve this problem, the heart beats faster. Physical activity causes as much as 10 times more blood to be sent to the muscles than when your body is at rest.

During exercise, some organs do not need as much blood as the skeletal muscles do. Less blood is sent to the kidneys and the digestive system so that more blood can go to the skeletal muscles, brain, heart, and lungs. This is like turning off certain water faucets in a house to allow more water to flow through other faucets.

Activity

Imagine that you are a scientist that has been chosen to explore the cardiovascular system. After being shrunk down to the size of a red blood cell, you board a miniature submarine and begin your travels. Describe where you go and what you see.

TRY at HOME

SECTION REVIEW

1. What is the function of the cardiovascular system?

2. What are the three kinds of blood vessels? Compare their functions.

3. **Identifying Relationships** How is the structure of capillaries related to their function?

What's Your Blood Type?

When a person loses a lot of blood, the person is given blood that has been donated from someone else. The person receiving the blood cannot be given blood from just anyone because people have different blood types. It's safe to mix some blood types, but mixing others causes a person's RBCs to clump together. The clumped cells may form blood clots, which block blood vessels, causing death.

Every person has one of the following blood types: A, B, AB, or O. Your blood type refers to the type of chemicals you have on the surface of your RBCs. These chemicals are called *antigens*. Type A blood has A antigens; type B has B antigens; and type AB has both A and B antigens. Type O blood has neither the A nor B antigen.

To Mix or Not to Mix Different blood types not only have different chemicals on their RBC surfaces but also may have different chemicals in their plasma, the liquid part of blood. These chemicals are *antibodies*. When antibodies bind to RBCs, they cause the RBCs to clump together.

As shown in **Figure 9,** the body makes antibodies against the antigens that are not on its own RBCs. For example, people with type B blood make A antibodies, which attack any blood cell with an A antigen on it. Therefore, people with type B blood can't be given type A or AB blood. Type O blood can be given to anyone because its RBCs don't have any A or B antigens on their surface. A person with type O blood is therefore said to be a *universal donor*. People with type AB blood are *universal recipients*, meaning they can receive any type of blood because they do not make any antibodies against A or B antigens.

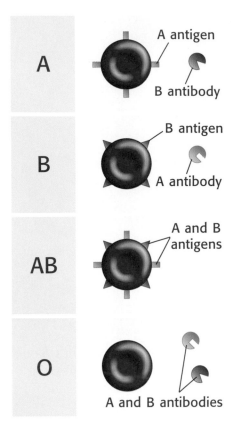

Figure 9 *This table shows which antigens and antibodies may be present in each blood type.*

Blood Delivery

A young woman is brought into the emergency room and needs a blood transfusion. Her blood type is AB. You call the blood bank to order AB blood, but you are told the bank is out of that type. What other type or types could the blood bank deliver for her transfusion?

Cardiovascular Problems

When something is wrong with a person's cardiovascular system, the person's health will be affected. Some cardiovascular problems involve the heart and the blood vessels, while other problems affect the blood. Cardiovascular problems can be caused by smoking, high levels of cholesterol in blood, stress, heredity, and other factors.

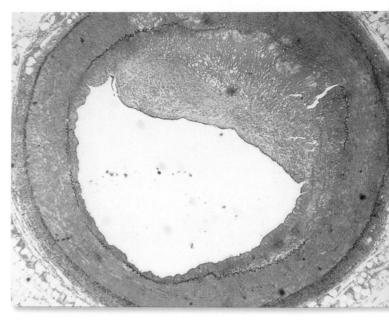

Atherosclerosis The leading cause of death in the United States is a cardiovascular disease called *atherosclerosis* (ATH uhr OH skluh ROH sis). Atherosclerosis occurs when fatty materials, such as cholesterol, build up on the inside of blood vessels. The fatty buildup causes the blood vessels to become narrower and less elastic. **Figure 10** shows how the pathway through a blood vessel can become clogged. When a major artery that supplies blood to the heart becomes blocked, a person has a heart attack, and part of the heart can die.

Figure 10 *Atherosclerosis is a common cardiovascular problem. Fatty deposits build up on the inside of blood vessels and block blood flow.*

A Point About Pressure Atherosclerosis also promotes *hypertension,* which is abnormally high blood pressure. Hypertension is dangerous because it overworks the heart and can weaken vessels and make them rupture. If a blood vessel in the brain becomes clogged or ruptures, certain parts of the brain will not receive oxygen and nutrients and may die. This is called a *stroke.*

MATH BREAK

The Beat Goes On

Your heart beats about 100,800 times per day. How many times does it beat per year?

SECTION REVIEW

1. Where does blood travel to and from during pulmonary circulation? during systemic circulation?

2. What happens to the oxygen level in blood as it moves through the lungs?

3. **Applying Concepts** Billy has type A blood.
   a. What kind of antigens does he have on his RBCs?
   b. What blood-type antibodies can Billy make?
   c. Which blood types could be given to Billy if he needed a transfusion?

internet**connect**

SC*L*INKS
NSTA

TOPIC: The Cardiovascular System, Cardiovascular Problems
GO TO: www.scilinks.org
*sci*LINKS NUMBER: HSTL555, HSTL560

Terms to Learn

lymphatic system   thymus
lymph   spleen
lymph nodes   tonsils

What You'll Do

♦ Discuss the functions of the lymphatic system.
♦ Identify the relationship between lymph and blood.
♦ Describe the organs of the lymphatic system.

The Lymphatic System

Your cardiovascular system is not the only circulatory system in your body. As blood flows through your cardiovascular system, fluid leaks out of the capillaries and mixes with the fluid that bathes your cells. Most of the fluid is reabsorbed by the capillaries, but some is not. To deal with this, your body's **lymphatic system** collects the excess fluid and returns it to your blood.

In addition to collecting the excess fluid surrounding your cells and returning it to your blood, your lymphatic system helps your body fight pathogens. Pathogens are microorganisms and viruses that make you sick.

Vessels of the Lymphatic System

The fluid collected by the lymphatic system is transported through vessels. The smallest vessels of the lymphatic system are *lymph capillaries.* From the spaces between cells, lymph capillaries absorb fluid and any particles too large to enter the blood capillaries. Some of these particles are dead cells or cells that are foreign to the body. The fluid and particles absorbed into lymph capillaries are called **lymph.**

As shown in **Figure 11,** lymph capillaries carry lymph into *lymphatic vessels,* which are larger vessels that have valves. Lymph is not pushed by a pump. Instead, the squeezing of skeletal muscles provides the force to move lymph through vessels, and valves help prevent backflow. Lymph travels through your lymphatic system and then drains into large neck veins of the cardiovascular system.

✓ **Self-Check**

How are the lymphatic system and the cardiovascular system similar? How are they different? *(See page 212 to check your answer.)*

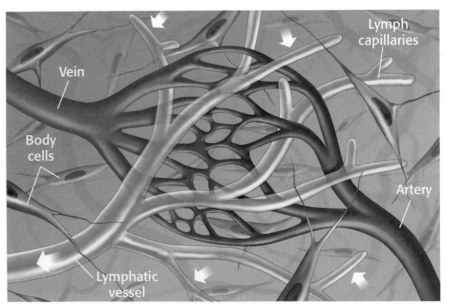

Figure 11 *The white arrows show the movement of lymph into lymph capillaries and through lymphatic vessels.*

Lymph capillaries

Vein

Body cells

Artery

Lymphatic vessel

Lymphatic Organs

In addition to vessels and capillaries, a variety of other organs are part of the lymphatic system, as shown in **Figure 12.**

Lymph Nodes As lymph travels through lymphatic vessels, it passes through lymph nodes. **Lymph nodes** are small bean-shaped organs where particles, such as pathogens or dead cells, are removed from the lymph.

Lymph nodes contain many white blood cells. Some of these cells engulf pathogens. Other WBCs produce chemicals that become attached to pathogens and mark them for destruction. When the body becomes infected with bacteria or viruses, the WBCs multiply and the nodes sometimes become swollen and painful.

Thymus Your **thymus,** which is located just above your heart, releases WBCs. The WBCs travel to other areas of the lymphatic system.

Spleen The largest lymph organ is your spleen, which is located in the upper left side of your abdomen. The **spleen** filters blood and, like the thymus, releases WBCs. When red blood cells are squeezed through the spleen's capillaries, the older and more fragile cells rupture. The RBCs are broken down, and some of their parts are reused. For this reason, the spleen can be thought of as the red-blood-cell recycling center.

Tonsils **Tonsils** are made up of groups of lymphatic tissue located at the back of your nasal cavity, on the inside of your throat, and at the back of your tongue. WBCs in the tonsils defend the body against infection. Tonsils sometimes become badly infected and must be removed.

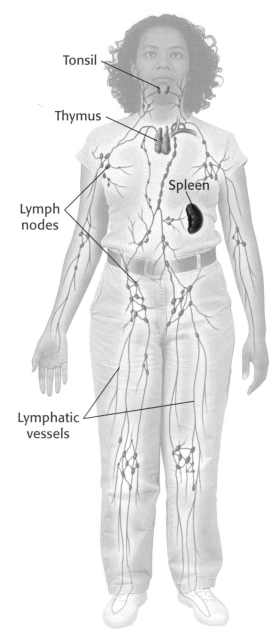

Figure 12 The Lymphatic System

SECTION REVIEW

1. What are the main functions of the lymphatic system?

2. Where does lymph go when it leaves the lymphatic system?

3. **Identifying Relationships** How are lymph nodes similar to the spleen?

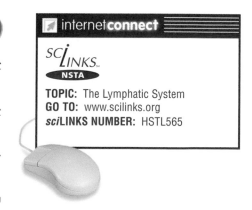

internet**connect**

SC*L*INKS

NSTA

TOPIC: The Lymphatic System
GO TO: www.scilinks.org
*sci*LINKS NUMBER: HSTL565

The Respiratory System

Breathing. You do it all the time. You're doing it right now. You hardly ever think about it, though, unless your ability to breathe is suddenly taken away. Then it becomes all too obvious that you must breathe in order to survive. Why is breathing important?

Out with the Bad Air; In with the Good

Your body needs a continuous supply of oxygen in order to obtain energy from the foods you eat. That's where breathing comes in handy. The air you breathe is a mixture of several gases. One of these gases is oxygen. When you breathe, your body takes in air and absorbs the oxygen. Then carbon dioxide from your body is added to the air, and the stale air is exhaled.

The words *breathing* and *respiration* are often thought to mean the same thing. However, breathing is only one part of respiration. **Respiration** is the entire process by which a body obtains and uses oxygen and gets rid of carbon dioxide and water. Respiration is divided into two parts: breathing, which involves inhaling and exhaling, and cellular respiration, which involves the chemical reactions that release energy from food.

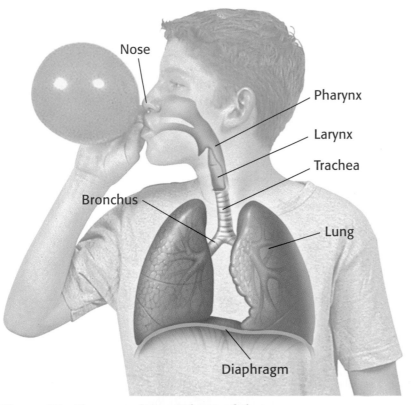

Nose
Pharynx
Larynx
Trachea
Bronchus
Lung
Diaphragm

Figure 13 *Air moves into and out of the body through the respiratory system.*

Breathing: Brought to You by Your Respiratory System

Breathing is made possible by the respiratory system. The **respiratory system** consists of the lungs, throat, and passageways that lead to the lungs. **Figure 13** shows the parts of the respiratory system.

Nose Your nose is the primary passageway into and out of the respiratory system. Air is inhaled through the nose, where it comes into contact with warm, moist surfaces. Air can also enter and leave through the mouth.

Pharynx From the nose, air flows into the **pharynx** (FER ingks), or throat. You can use a mirror to see the walls of your pharynx behind your tongue. In addition to air, food and drink also travel through the pharynx on the way to the stomach. The pharynx branches into two tubes. One leads to the stomach and is called the esophagus. The other leads to the lungs and is called the larynx.

Larynx Tilt your head up slightly, and rub a finger up and down the front of your throat. The ridges you feel are the outside of the larynx (LER ingks). The **larynx,** or voice box, contains the vocal cords. The vocal cords are a pair of elastic bands that are stretched across the opening of the larynx. Muscles attached to the larynx control how much the vocal cords are stretched. When air flows between the vocal cords, they vibrate and make sound.

Trachea The larynx guards the entrance to a large tube called the **trachea** (TRAY kee uh), or windpipe. The trachea is the passageway for air traveling from the larynx to the lungs.

Bronchi The trachea splits into two tubes called **bronchi** (BRAHNG kie) (singular, *bronchus*). One bronchus goes to each lung and branches into thousands of tiny tubes called *bronchioles*.

Lungs Your body has two large spongelike lungs. In the lungs, each bronchiole branches to form thousands of tiny sacs called **alveoli** (singular, *alveolus*). Capillaries surround each alveolus. **Figure 14** shows the arrangement of the tubes in the respiratory system.

Chemistry
CONNECTION

When people who live at low elevations travel to the mountains, they usually find that they have difficulty exerting themselves. This is because the concentration of oxygen in the air at high elevations is lower than that at low elevations. Until they become used to the change, people have to take more breaths to supply their bodies with the oxygen they need.

Figure 14 *Inside your lungs, the bronchi branch into bronchioles. The bronchioles lead to tiny sacs called alveoli.*

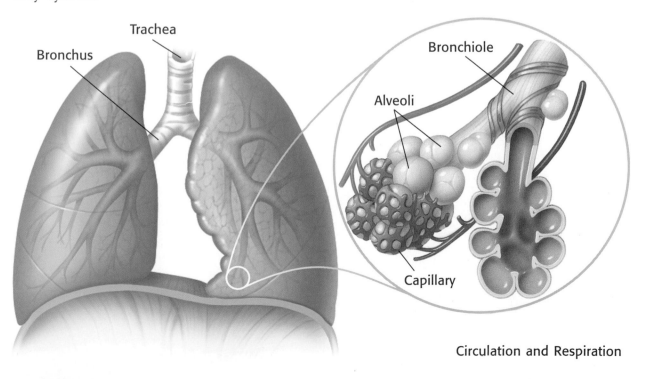

Trachea
Bronchus
Bronchiole
Alveoli
Capillary

How Do You Breathe?

When you breathe, air is sucked in or forced out of your lungs. However, your lungs do not contain muscles that force air in and out. Instead, breathing is done by rib muscles and the *diaphragm,* a dome-shaped muscle underneath the lungs. When the diaphragm contracts and moves down, it increases the chest cavity's volume. At the same time, some of your rib muscles contract and lift your rib cage, causing it to expand. Air is sucked in.

What Happens to the Oxygen? When oxygen has been absorbed by red blood cells, it is transported through the body by the cardiovascular system. Oxygen diffuses inside cells, where it is used in an important chemical reaction known as cellular respiration. During *cellular respiration,* oxygen is used to release energy stored in molecules of carbohydrates, fats, and proteins. When the molecules are broken apart during the reaction, energy is released along with carbon dioxide and water. The carbon dioxide and water leave the cell and return to the bloodstream. The carbon dioxide is carried to the lungs and exhaled. **Figure 15** shows how breathing and blood circulation are related.

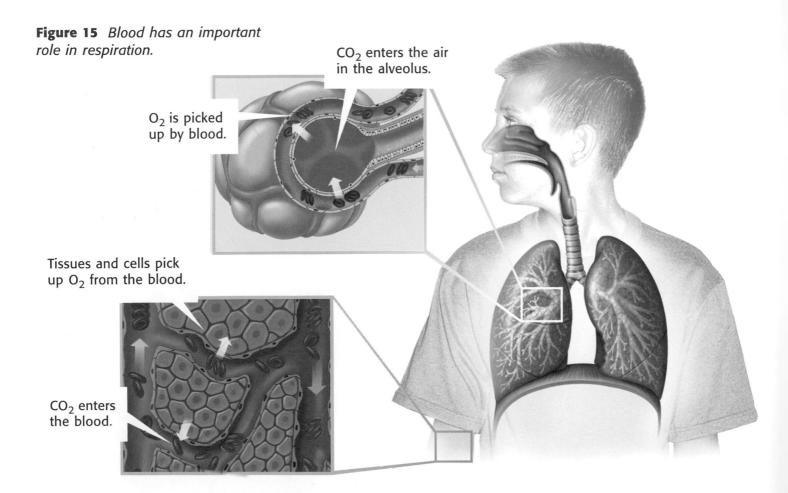

Figure 15 *Blood has an important role in respiration.*

CO_2 enters the air in the alveolus.

O_2 is picked up by blood.

Tissues and cells pick up O_2 from the blood.

CO_2 enters the blood.

Respiratory Disorders

Millions of people suffer from respiratory disorders. There are many types of respiratory disorders, including asthma, bronchitis, pneumonia, and emphysema.

In *asthma,* irritants cause tissue around the bronchioles to constrict and secrete large amounts of mucus. As the bronchiole tubes get narrower, the person has difficulty breathing. *Bronchitis* can develop when something irritates the lining of the bronchioles. *Pneumonia* is caused by bacteria or viruses that grow inside the bronchioles and alveoli and cause them to become inflamed and filled with fluid. If the alveoli are filled with too much fluid, the person may suffocate.

The Hazards of Smoking You probably already know that smoking cigarettes is bad for your health. In fact, smoking is the leading cause of cardiovascular diseases and lung diseases, such as *emphysema* and *lung cancer.* People with emphysema have trouble getting the oxygen they need because their lung tissue erodes away, as shown in **Figure 16.**

Why Do People Snore?

Get a **15 cm² sheet of wax paper.** Hum your favorite song. Then take the wax paper, press it against your lips, and hum the song again. Now answer the following questions:

1. How was your humming different when wax paper was pressed to your mouth?

2. Use your observations to guess what might cause snoring.

TRY at HOME

Figure 16 The photo on the left shows healthy lungs. The photo on the right shows the lungs of a person who had emphysema.

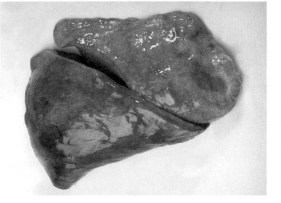

SECTION REVIEW

1. Describe the path that air travels as it moves through the respiratory system.

2. What is the difference between respiration and cellular respiration?

3. **Identifying Relationships** How is the function of the respiratory system related to that of the cardiovascular system?

internet**connect**

SC*LINKS*
NSTA

TOPIC: The Respiratory System, Respiratory Disorders
GO TO: www.scilinks.org
*sci***LINKS NUMBER:** HSTL570, HSTL575

Making Models Lab

Build a Lung

You have learned that when you breathe, you actually pull air into your lungs because your diaphragm muscle causes your chest to expand. You can see this is true by placing your hands on your ribs and inhaling slowly. Did you feel your chest expand?

In this activity, you will build a model of a lung, using some common materials. You will see how the diaphragm muscle works to inflate your lungs. Refer to the diagrams as you construct your model.

MATERIALS

- small balloon
- plastic drinking straw
- 2 rubber bands
- golf-ball-sized piece of clay
- metric ruler
- top half of a 2 L bottle
- small plastic trash bag
- transparent tape

Procedure

1 Attach the balloon to the end of the straw with a rubber band. Make a hole through the clay, and insert the other end of the straw through the hole. Be sure at least 8 cm of the straw extends beyond the clay. Squeeze the ball of clay gently to seal the clay around the straw.

2 Insert the balloon end of the straw into the neck of the bottle. Use the ball of clay to seal the straw and balloon into the bottle.

3 Place the trash bag over the cut end of the bottle, and secure it with one rubber band. Reinforce the seal with tape. Gather the excess material of the bag into your hand, and press toward the inside of the bottle slightly. This will push the excess air out of the bottle. Tape the bag to the bottle with the bag in this position.

Analysis

4 How can you make your model "lung" inflate?

5 What do the balloon, the plastic bag, and the straw represent in your model?

6 Using your model, demonstrate how air enters the lung and how air exits the lung.

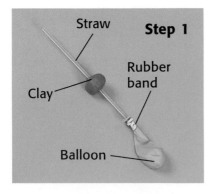
Straw
Clay
Rubber band
Balloon
Step 1

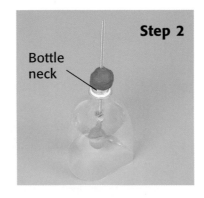
Bottle neck
Step 2

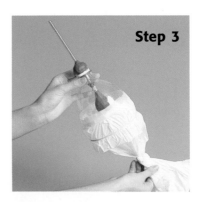
Step 3

Skill Builder Lab

Carbon Dioxide Breath

Plants take in carbon dioxide and give off oxygen as a byproduct of photosynthesis. Animals, including you, use this oxygen and release carbon dioxide as a byproduct of respiration.

In this activity, you will explore your own carbon dioxide exhalation. Phenol red turns yellow in the presence of carbon dioxide. You will use it to detect carbon dioxide in your breath.

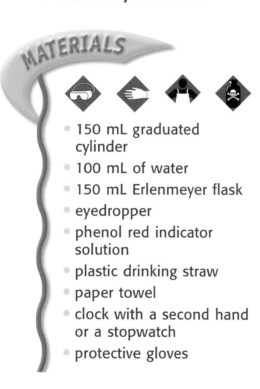

MATERIALS

- 150 mL graduated cylinder
- 100 mL of water
- 150 mL Erlenmeyer flask
- eyedropper
- phenol red indicator solution
- plastic drinking straw
- paper towel
- clock with a second hand or a stopwatch
- protective gloves

Procedure

1 Place 100 mL of water into a 150 mL flask. Using an eyedropper, carefully place four drops of phenol red indicator solution into the water. The water should turn orange.

2 Place a plastic drinking straw into the solution of phenol red and water. Drape a paper towel over the beaker to prevent splashing. Carefully blow through the straw into the solution.
Caution: Do not inhale through the straw. Do not drink the solution, and do not share a straw with anyone.

3 Have your partner time how long it takes for the solution to change color. Begin timing when you start blowing. Record the time in your Science-Log. What color does the solution become?

Analysis

4 Compare your data with those of your classmates. What was the longest length of time it took to see a color change? the shortest? How do you account for the difference?

5 Is there a relationship between the time it takes to change the solution from orange to yellow and the person's physical characteristics, such as gender or whether the tester has an athletic build?

Going Further

Do jumping jacks or sit-ups for three minutes, and then repeat the experiment. Did the timing change? Describe and explain any change.

Chapter Highlights

SECTION 1

Vocabulary

cardiovascular system *(p. 30)*

blood *(p. 30)*

arteries *(p. 33)*

capillaries *(p. 33)*

veins *(p. 33)*

pulmonary circulation *(p. 34)*

systemic circulation *(p. 34)*

blood pressure *(p. 35)*

Section Notes

- The cardiovascular system delivers oxygen and nutrients to the body's cells, takes away the cells' waste products, and helps the body stay healthy. The cardiovascular system is made up of blood, the heart, and blood vessels.

- Blood is a connective tissue made of plasma, red blood cells, white blood cells, and platelets. The heart is a muscular organ that pumps blood through blood vessels.

- Blood moves away from the heart through arteries and then enters capillaries. After leaving capillaries, blood is carried back to the heart through veins.

- In pulmonary circulation, blood vessels carry blood from the heart to the lungs and back to the heart. In systemic circulation, blood flows from the heart to the rest of the body and then back to the heart.

- People have different blood types. Blood type is determined by the presence of certain chemicals on red blood cells.

SECTION 2

Vocabulary

lymphatic system *(p. 38)*

lymph *(p. 38)*

lymph nodes *(p. 39)*

thymus *(p. 39)*

spleen *(p. 39)*

tonsils *(p. 39)*

Section Notes

- The lymphatic system returns excess fluid to the cardiovascular system and helps the body fight infections.

- The lymphatic system includes lymph, lymph capillaries, lymphatic vessels, lymph nodes, the spleen, tonsils, and the thymus.

☑ Skills Check

Math Concepts

A CONTINUOUS BEAT Your heart beats about 100,800 times per day. That means that your heart beats about 4,200 times every hour.

100,800 beats ÷ 24 hours = 4,200 beats

That also means that your heart beats about 70 times every minute.

4,200 beats ÷ 60 minutes = 70 beats

Visual Understanding

AIR PASSAGEWAYS Take another look at Figure 13 on page 40. With your finger, trace the path air takes to reach the lungs. As you do this, reconsider what roles the nose, pharynx, trachea, bronchi, lungs, and diaphragm play in respiration.

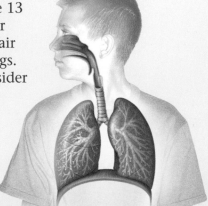

Vocabulary

respiration *(p. 40)*

respiratory system *(p. 40)*

pharynx *(p. 41)*

larynx *(p. 41)*

trachea *(p. 41)*

bronchi *(p. 41)*

alveoli *(p. 41)*

Section Notes

- The respiratory system moves air into and out of the body. The respiratory system includes the nose, the mouth, the pharynx, the larynx, the trachea, and the lungs.

- Air enters the lungs through bronchi and travels to the alveoli, which are gas-filled sacs surrounded by capillaries of the cardiovascular system.

- The blood in the capillaries of the lungs absorbs oxygen and releases carbon dioxide. The carbon dioxide is exhaled. The oxygen is carried by the blood to the heart and then on to the cells of the body.

- The body's cells must have oxygen to carry out cellular respiration. Cellular respiration is a chemical process that releases the energy in carbohydrates, fats, and proteins and makes the energy available to the cells.

- Inhaling and exhaling are caused by the contraction and relaxation of the diaphragm and the muscles of the rib cage.

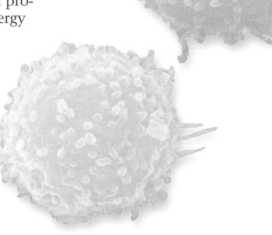

 internet **connect**

 GO TO: go.hrw.com

Visit the **HRW** Web site for a variety of learning tools related to this chapter. Just type in the keyword:

KEYWORD: HSTBD2

 **SCI** **LINKS** SM **NSTA** **GO TO:** www.scilinks.org

Visit the **National Science Teachers Association** on-line Web site for Internet resources related to this chapter. Just type in the *sci*LINKS number for more information about the topic:

TOPIC: The Cardiovascular System	*sci*LINKS NUMBER: HSTL555
TOPIC: Cardiovascular Problems	*sci*LINKS NUMBER: HSTL560
TOPIC: The Lymphatic System	*sci*LINKS NUMBER: HSTL565
TOPIC: The Respiratory System	*sci*LINKS NUMBER: HSTL570
TOPIC: Respiratory Disorders	*sci*LINKS NUMBER: HSTL575

Chapter Review

To complete the following sentences, choose the correct term from each pair of terms listed below:

1. Oxygen is delivered to the cells of the body by __?__. (*white blood cells* or *red blood cells*)

2. Blood is carried away from the heart in __?__. (*arteries* or *veins*)

3. __?__ carries nutrients to the body's cells. (*Lymph* or *Blood*)

4. The __?__ contains the vocal cords. (*trachea* or *larynx*)

5. The pathway of air through the respiratory system ends at the tiny sacs called __?__. (*alveoli* or *bronchi*)

UNDERSTANDING CONCEPTS

Multiple Choice

6. Blood from the lungs enters the heart at the
   a. left ventricle.
   b. left atrium.
   c. right atrium.
   d. right ventricle.

7. Blood cells are made
   a. in the heart.
   b. from plasma.
   c. from lymph.
   d. in the bones.

8. Which of the following is not part of the lymphatic system?
   a. trachea
   b. lymph node
   c. thymus
   d. spleen

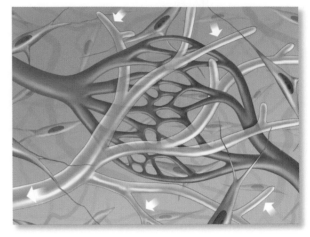

9. Alveoli are surrounded by
   a. veins.
   b. muscles.
   c. capillaries.
   d. lymph nodes.

10. What prevents blood from flowing backward in veins?
    a. platelets
    b. valves
    c. muscles
    d. cartilage

11. Air moves into the lungs when the diaphragm muscle
    a. contracts and moves down.
    b. contracts and moves up.
    c. relaxes and moves down.
    d. relaxes and moves up.

Short Answer

12. What is the difference between pulmonary circulation and systemic circulation in the cardiovascular system?

13. Walton has a blood pressure of 110/65. What do the two numbers mean?

14. What body process produces the carbon dioxide you exhale?

Concept Map

15. Use the following terms to create a concept map: blood, oxygen, alveoli, capillaries, carbon dioxide.

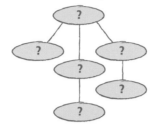

CRITICAL THINKING AND PROBLEM SOLVING

Write one or two sentences to answer the following questions:

16. Why do you think there are hairs in your nose?

17. When a person is not feeling well, sometimes a doctor will examine samples of the person's blood to see how many white blood cells are present. Why would this information be useful?

18. How is the function of the lymphatic system related to the function of the cardiovascular system?

MATH IN SCIENCE

19. After a person donates blood, the blood is stored in one-pint bags until it is needed for a transfusion. A healthy person normally has 5 million RBCs in each cubic millimeter ($1\ mm^3$) of blood.
    a. How many RBCs are there in 1 mL of blood? One milliliter is equal to $1\ cm^3$ and to $1{,}000\ mm^3$.
    b. How many RBCs are there in 1 pt? One pint is equal to 473 mL.

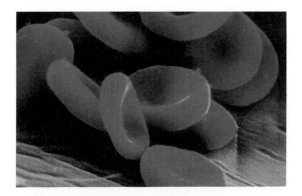

INTERPRETING GRAPHICS

The diagram below shows how the human heart would look in cross section. Examine the diagram, and then answer the questions that follow:

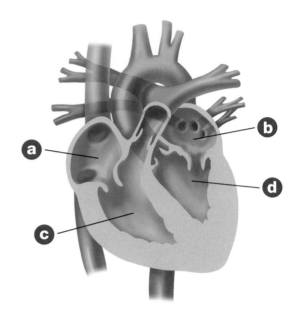

20. Which letter identifies the chamber that receives blood from systemic circulation? What is this chamber's name?

21. Which letter identifies the chamber that receives blood from the lungs? What is this chamber's name?

22. Which letter identifies the chamber that pumps blood to the lungs? What is this chamber's name?

Reading Check-up Take a minute to review your answers to the Pre-Reading Questions found at the bottom of page 28. Have your answers changed? If necessary, revise your answers based on what you have learned since you began this chapter.

Circulation and Respiration **49**

CATCHING A LIGHT SNEEZE

Do you sneeze when you come out of a dark movie theater into bright sunlight? If not, look around you next time. Chances are several people will sneeze.

Reflex Gone Wrong

For some reason, about one in five people sneeze when they step from a dimly lit area into a brightly lit area. In fact, some may sneeze a dozen times or more! Fortunately, the sneezing usually stops after a few times. This reaction is called a *photic sneeze reflex.* No one knows for certain why it happens.

Normal sneezing is a reflex, which means you do it without thinking about it. Most people sneeze when something tickles the inside of their nose. They sneeze, and moving air pushes the tickling intruder out. For instance, if you get dust in your nose, sneezing pushes the dust out. In the case of people with the photic sneeze, it's a reflex gone wrong.

ACHOO!

A few years ago, some geneticists studied the photic sneeze reflex. They named it the Autosomal Dominant Compelling Helio-ophthalmic Outburst syndrome, or the ACHOO syndrome. Scientists know that the ACHOO syndrome runs in families. So the photic sneeze can be passed from parent to child. Sometimes even the number of times in a row that each person sneezes is the same throughout a family.

Possible Answers

Some scientists have offered a possible explanation for the ACHOO syndrome. First, everyone's pupils contract when they

▲ *Do you sneeze when you see bright light after exiting a dark room?*

encounter bright light. And the nerves from the eyes are right next to the nerves from the nose. Thus, people with the ACHOO syndrome may have their wires slightly crossed: bright light triggers the pupil reflex, and it also triggers the sneeze reflex!

Sneeze Fest

Sunlight is not the only strange trigger for sudden sneezes. Some people sneeze when they rub the inner corner of their eye. Others sneeze when tweezing their eyebrows or brushing their hair. In rare individuals, even eating too much has been known to cause sneezing fits!

Research the Facts

▶ Yawning is also a reflex. Do some research to find out why we yawn.

Health

Goats to the Rescue

They're called transgenic (tranz JEHN ik) goats because their cells contain a human gene. They look just like any other goats, but because of their human gene they produce a chemical that can save lives.

Lifesaving Genes

Heart attacks are the number one cause of death in the United States. Many heart attacks are triggered when large blood clots interfere with the flow of blood to the heart. Human blood cells produce a chemical called *tissue plasminogen activator* (TPA) that dissolves small blood clots. If TPA is given to a person having a heart attack, it can often dissolve the blood clot, stop the attack, and save the person's life. But TPA is difficult to produce in large quantities in the laboratory. This is where the goats come in. Researchers at Tufts University, in Grafton, Massachusetts, have genetically engineered goats to produce this lifesaving drug.

▲ *A scientist at Tufts University injects human TPA genes into fertilized goat eggs.*

Hybrid Goats

Producing transgenic goats is a complicated process. First, fertilized eggs are surgically removed from normal female goats. The eggs are then injected with hybrid genes that consist of human TPA genes "spliced" into genes from the mammary glands of a goat. Finally, the altered eggs are surgically implanted into female goats, where they develop into young goats, or kids. Some of the kids actually carry the hybrid gene. When the hybrid kids mature, the females' milk contains TPA. Technicians then separate the TPA from the goats' milk for use in heart-attack victims.

The Research Continues

Transgenic research in farm animals such as goats, sheep, cows, and pigs may someday produce drugs faster, cheaper, and in greater quantities than are possible using current methods. The way we view the barnyard may never be the same.

Find Out for Yourself

▶ Using chemicals produced by transgenic animals is just one of many gene therapies. Do some research to find out more about gene therapy, how it is used, and how it may be used in the future.

The Digestive and Urinary Systems

Pre-Reading Questions

1. What happens to food when it reaches your stomach?

2. What does your urinary system have in common with your skin?

AHHH!!!

You can easily see the stream of water going into this girl's mouth. What you cannot see is where that refreshing gulp of water goes after she swallows it. How do water and food travel through the body and get absorbed for use by the cells? What are the body systems for getting rid of excess water or wastes? In this chapter, you will learn about the organs of the digestive and urinary systems and what happens inside your body to the food and liquids you eat and drink.

START-UP Activity

CHANGING FOODS

During digestion, the stomach squeezes and relaxes as food passes through it. You can see the role this squeezing motion plays in the digestion of food by using a plastic bag to model your stomach.

Procedure

1. Add **200 mL of flour** and **100 mL of water** to a **resealable plastic bag.** Mix **100 mL of vegetable oil** with the flour and water.

2. Seal the plastic bag, and shake it until the flour, water, and oil are well mixed.

3. Remove as much air from the bag as possible, and reseal the bag carefully.

4. Squeeze the bag with your hands for 5 minutes. Record your observations in your ScienceLog. Be careful to keep the bag sealed.

Analysis

5. Describe the mixture before and after you kneaded the bag.

6. How might the changes you saw in the mixture relate to what happens to food you eat?

7. Do you think this is a good model of how your stomach works? Why or why not?

The Digestive System

Terms to Learn

digestive system   pancreas
esophagus   liver
stomach   gallbladder
small intestine   large intestine

What You'll Do

◆ Describe the parts and functions of the digestive system.
◆ Compare mechanical digestion with chemical digestion.
◆ Describe some disorders of the digestive system.

It's your last class before lunch, and you're starving! You are so hungry you can hardly concentrate. Finally the bell rings and you get to eat your peanut butter and jelly sandwich. Yum!

You feel hungry because your brain receives signals that your cells need energy, but eating is only the beginning of the story. Your body must change a meal into substances it can use. Your **digestive system** is a group of organs that work together to digest food so that it can be used by the body.

Digestive System at a Glance

The most obvious part of your digestive system is the *digestive tract,* a series of tubelike organs that are joined end to end. The digestive tract includes your mouth, throat, esophagus, stomach, small intestine, large intestine, rectum, and anus. The human digestive tract may be more than 9 m long! The food you eat is digested as it passes through these organs. The liver, gallbladder, pancreas, and salivary glands are also part of the digestive system because they secrete substances that are used in digestion. The digestive system is shown in **Figure 1.**

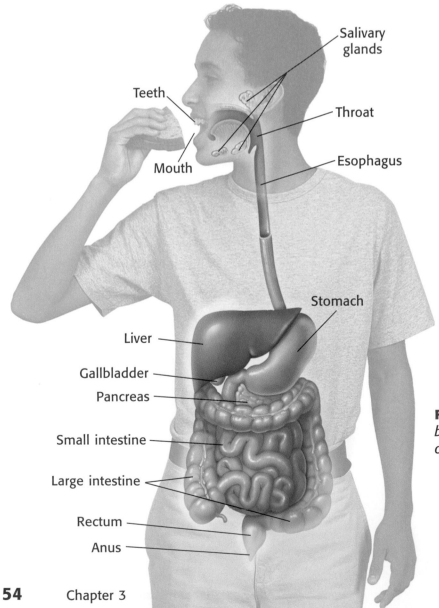

Figure 1 *The digestive tract is basically a long tube with an opening at each end.*

The Journey of a Sandwich

Digestion is the process of breaking down food, such as a peanut butter and jelly sandwich, into a form that can pass from the digestive tract into the bloodstream. There are two types of digestion—mechanical and chemical. The breaking, crushing, and mashing of food is called *mechanical digestion.* In *chemical digestion,* large molecules are broken down into nutrients. Nutrients are substances in food that the body needs for normal growth, maintenance, and repair.

Three major types of nutrients—carbohydrates, proteins, and fats—make up most of what you eat. In fact, a peanut butter and jelly sandwich contains all three of these nutrients. Special substances called *enzymes* break some nutrients into smaller particles that the body can use. For example, proteins are chains of smaller molecules called amino acids. Proteins are too large to be absorbed into the bloodstream, but enzymes chop up the chain into amino acids. These amino acids are small enough to pass into the bloodstream. This process is illustrated in **Figure 2.**

Quick Lab

Break It Up!

1. Drop **one piece of hard candy** into a **clear plastic cup of water.**

2. Wrap an **identical candy** in a **towel,** and crush it with a **hammer.** Drop the crushed candy into a **second clear cup of water.**

3. The next day, examine both cups. What is different about the two candies?

4. What part of digestion is represented by breaking the hard candy?

5. How does chewing your food help the process of digestion?

Figure 2 *Enzymes in the stomach and small intestine break down proteins.*

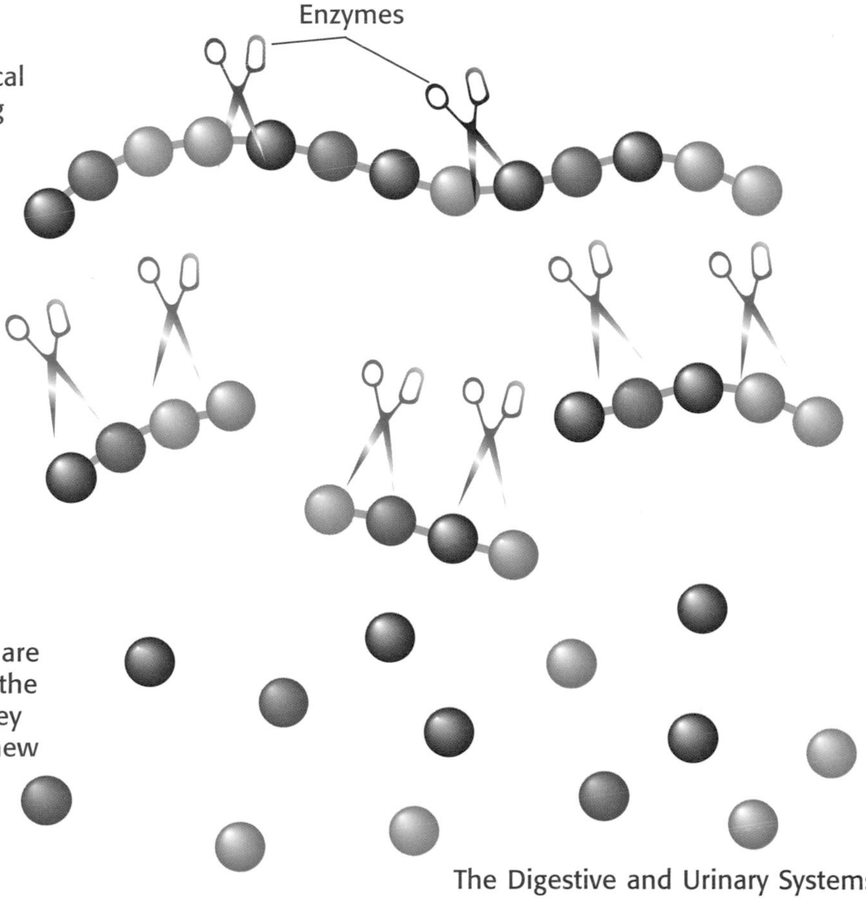

Enzymes

❶ Enzymes act as chemical scissors to cut the long chains of amino acids into small chains.

❷ The small chains are then split by other enzymes.

❸ Individual amino acids are small enough to enter the bloodstream, where they can be used to make new proteins.

The Digestive and Urinary Systems **55**

Digestion Begins in the Mouth

Why is chewing so important? There are two reasons. First, chewing creates small, slippery pieces of food that are easier to swallow than big, dry pieces. Second, small pieces of food are easier to digest.

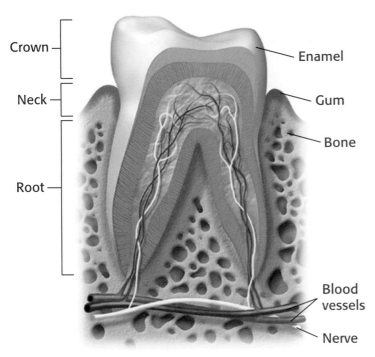

Figure 3 *The crown of a tooth, such as this molar, is visible above the gum line. The root is below the gum line.*

Through the Teeth Teeth are very important organs for mechanical digestion. With the help of strong muscles and your jaw bones, teeth are able to break and grind food. The outermost layer of a tooth, the *enamel,* is the hardest material in the body. Enamel protects nerves and softer material inside the tooth. **Figure 3** shows the major parts of the tooth.

Have you ever noticed that your teeth have different shapes? Look at **Figure 4** to locate the different kinds of teeth. The *molars* in the back are well suited for grinding food. The *premolars* are perfect for mashing food. The sharp teeth at the front of your mouth, the *incisors* and *canines,* are for shredding food.

And Over the Gums As you chew, the food gets mixed with a liquid called *saliva.* Saliva is made in salivary glands located in and around the mouth. Saliva contains an enzyme that begins the chemical digestion of carbohydrates. Saliva turns complex carbohydrates into simple sugars.

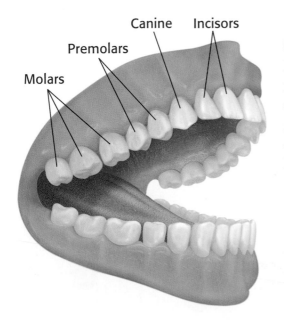

Look Out Stomach, Here It Comes! Once the food has been reduced to a soft mush, the tongue pushes it into the throat, which leads to a long, straight tube called the **esophagus** (i SAWF uh guhs). The esophagus squeezes the mass of food with rhythmic muscle contractions called *peristalsis* (PER uh STAHL sis). Peristalsis forces the food into the stomach.

Figure 4 *Most adults have 32 permanent teeth. Each type of permanent tooth has a different function in breaking up food before it is swallowed.*

The Stomach's Harsh Environment

The **stomach** is a muscular, baglike organ attached to the lower end of the esophagus. It is pictured in **Figure 5.** The stomach continues the physical digestion of your meal by squeezing its contents with muscular contractions. While all this squeezing is going on, tiny glands in the stomach produce enzymes and acid. These work together to break food into nutrients. Stomach acid also kills most bacteria that you might swallow with your food. After a few hours of combined physical and chemical action, your peanut butter and jelly sandwich has been reduced to a soupy mixture called *chyme* (kiem).

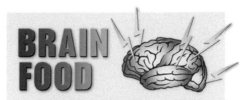

A thick substance called mucus covers the stomach's lining and offers some protection from its harsh environment. However, the acids still damage the lining, and the entire lining must be replaced every few days.

Figure 5 *The stomach grinds and mixes food for hours before it releases the mixture into the small intestine.*

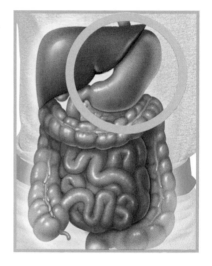
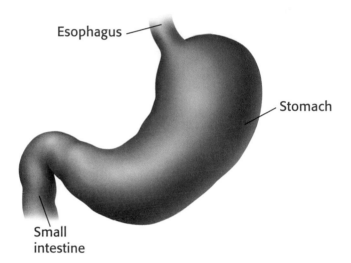

Esophagus

Stomach

Small intestine

Doorway to the Small Intestine The chyme is slowly released into the small intestine through a small ring of muscle that works like a valve. This valve keeps food in the stomach until it has been thoroughly mixed with digestive fluids. Then the valve opens and closes, letting a small amount of chyme squirt into the small intestine each time. Releasing chyme slowly from the stomach gives the intestine more time to mix the chyme with fluids from the liver and pancreas.

SECTION REVIEW

1. What is the difference between mechanical digestion and chemical digestion?

2. **Inferring Conclusions** Give two reasons why the following statement is true: Digestion begins in the mouth.

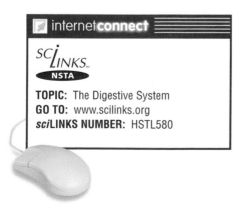

internet**connect**

SC*L*INKS
NSTA

TOPIC: The Digestive System
GO TO: www.scilinks.org
*sci*LINKS NUMBER: HSTL580

The Gigantic Small Intestine?

The **small intestine** is a muscular tube that is about 2.5 cm in diameter. Other than its diameter, it's really not that small at all. In fact, if you stretched it out, it would be longer than you are tall—about 6 m!

Villi If you flattened out the surface of the small intestine, it would be larger than a tennis court! How is this possible? The inside wall of the small intestine is covered with fingerlike projections called *villi*, shown in **Figure 6.** The villi are covered with tiny nutrient-absorbing cells. Because the villi extend into the chyme, these cells have greater exposure to nutrients. Once absorbed, the nutrients enter the bloodstream.

Most chemical digestion takes place in the small intestine. Chyme from the stomach moves very slowly through the small intestine by peristalsis. Proteins, carbohydrates, and fats in the chyme are digested with the help of enzymes produced in the small intestine and the pancreas.

Still hungry for news about digestion? Enzymes can help you with that steak, you know. Check it out on page 180 of the LabBook.

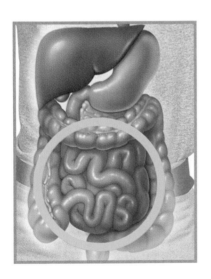

Figure 6 *The highly folded lining of the small intestine has many fingerlike extensions called villi.*

Villi are covered with nutrient-absorbing cells that pass nutrients on to the bloodstream.

The Pancreas

The **pancreas** is a fish-shaped organ located between the stomach and small intestine. It makes pancreatic juice that flows into the small intestine. This juice contains digestive enzymes and bicarbonate that neutralizes the acid in chyme. Without bicarbonate, acids would damage the lining of the intestine and prevent enzymes from doing their work. The pancreas also functions as a part of the endocrine system, making hormones that regulate blood sugar. The pancreas is shown in **Figure 7** on the next page.

The Liver and Gallbladder

The **liver** is a large reddish brown organ that helps with diges-tion. A human liver can be as large as a football. Your liver is located toward your right side, slightly higher than your stomach, as shown in **Figure 7.** Here are a few of the liver's important jobs:

- Your liver makes a green liquid called *bile* that is used in fat digestion
- Your liver stores nutrients
- Your liver breaks down toxic substances in the blood
- Your liver makes cholesterol for cell membranes

Bile Breaks Up Fat Although bile is made by the liver, it is temporarily stored in a small baglike organ called the **gallbladder,** shown in Figure 7. Bile is squeezed from the gall-bladder into the small intestine, where it breaks up large fat droplets into very small droplets. This physical process allows more fat molecules to be exposed to digestive enzymes.

Storing Nutrients and Protecting the Body After nutrients are broken down, they are absorbed into the bloodstream and carried through the body. Nutrients that are not needed right away are stored in the liver. The liver then releases the stored nutrients into the bloodstream as needed. The liver also captures and detoxifies many substances in the body. For instance, it produces enzymes that break down alcohol and many other drugs.

BRAIN FOOD

If three-fourths of the liver were removed, the rest would go on working and would eventually grow to replace the part that was removed.

✓ Self-Check

Is bile used for chemi-cal or mechanical digestion? Explain. *(See page 212 to check your answer.)*

Figure 7 *The liver, gallbladder, and pancreas are linked to the small intestine, but food does not move through them.*

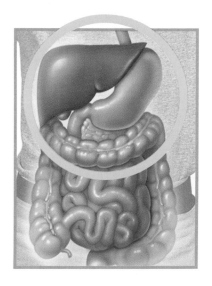

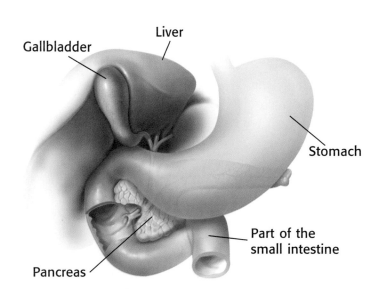

Gallbladder

Liver

Stomach

Part of the small intestine

Pancreas

The End of the Line

Whatever can't be absorbed into the blood gets pushed into the large intestine. The **large intestine** is the organ of the digestive system that stores, compacts, and then eliminates indigestible material from the body. The large intestine, shown in **Figure 8,** is called "large" because it has a larger diameter than the small intestine. It is about 1.5 m long, and its diameter is about 7.5 cm.

In the Large Intestine Undigested material enters the large intestine as a soupy mixture. The large intestine reabsorbs most of the water in the mixture, changing the liquid into a solid mass called *feces* or *stool*.

Whole grains, fruits, and vegetables contain a carbohydrate, called cellulose, that humans cannot digest. We commonly refer to this material as fiber. Fiber keeps the stool soft and keeps things moving through the large intestine.

A Way Out The *rectum* is the last section of the large intestine. It stores feces until they can be expelled. Feces pass to the outside through an opening called the *anus*. It has taken your sandwich about 24 hours to make this journey.

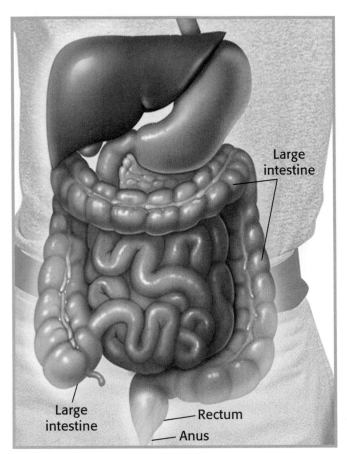

Large intestine

Large intestine — Rectum — Anus

Figure 8 *The large intestine is the final organ of digestion.*

Environment
C O N N E C T I O N

Feces and other human wastes contain microorganisms and other substances that can contaminate drinking water. Every time you flush a toilet, the water and wastes go through the sewer to a sewage treatment plant. Here the disease-causing microorganisms are removed, and the clean water is released back to rivers, lakes, and streams.

Problems in the Digestive System

Disorders of the digestive system are frequently related to eating behaviors. However, digestive problems can also be caused by diseases. Some common digestive disorders are described below.

Heartburn The stomach is blocked off at either end by bands of muscle called *sphincters* (SFINGK tuhrz). Occasionally, backflow of chyme from the stomach to the esophagus causes a burning pain in the chest called heartburn. Eating too much, eating right before going to bed, and eating very acidic foods sometimes cause heartburn.

Constipation and Diarrhea When the body does not get enough fiber, water, or exercise, the contents of the large intestine can become dry. Bowel movements become difficult and less frequent. This condition is called *constipation*. When bowel movements are frequent and watery, the condition is called *diarrhea*. Diarrhea occurs when too little water is removed from digested food in the large intestine. Diarrhea may cause dehydration and is especially dangerous for infants and small children, such as the girl in **Figure 9**.

Figure 9 *This child is being given fluids to replace those lost to diarrhea.*

Colon Cancer *Colon cancer* is a serious disease of the digestive tract that can lead to death. The colon is the long tubular portion of the large intestine. When certain colon cells divide uncontrollably, a tumor forms. Tumors interfere with the normal functioning of organs. Cells from a tumor can also break away and start tumors in other areas in the body. Colon cancer can often be treated and cured if detected early.

Gastric Ulcer An open sore in the stomach lining is called a *gastric ulcer*. **Figure 10** shows stomach tissue from a gastric ulcer. Gastric ulcers are often caused by bacteria and can be treated successfully with antibiotics. A high-fat diet, smoking, caffeine, and alcohol may make this condition worse.

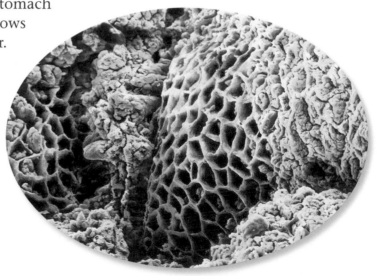

Figure 10 *This stomach lining has openings, seen in red, that indicate a gastric ulcer.*

SECTION REVIEW

1. What happens to the food that you eat when it gets to the stomach?

2. Describe the roles of the liver, the gallbladder, and the pancreas in digestion.

3. **Analyzing Relationships** How would the inability to make saliva affect digestion?

internet**connect**

SCI*LINKS*
NSTA

TOPIC: Problems in the Digestive System
GO TO: www.scilinks.org
*sci*LINKS NUMBER: HSTL585

The Urinary System

Terms to Learn

urinary system   urine
kidney   urinary bladder
nephron

What You'll Do

◆ Describe the parts and functions of the urinary system.
◆ Explain how the kidneys filter blood.
◆ Describe some disorders of the urinary system.

As your body performs the chemical activities that keep you alive, waste products such as carbon dioxide and nitrogen are produced. Your body has to get rid of these waste products in order to stay healthy. *Excretion* is the process of removing wastes and excess products from the body. Three of your body systems are involved in excretion: your skin releases waste products and water when you sweat, your lungs expel carbon dioxide and water when you exhale, and the **urinary system** removes waste products from your blood. Notice that the digestive system is not involved in excretion. The term *excretion* is used only when substances must pass through a membrane in order to leave the body.

Cleaning the Blood

As blood travels through the tissues, it collects all of the waste products produced by the body's cells. Your blood is like a supply train that comes into a town to drop off supplies and take away garbage. The train has to find a way to get rid of the garbage before it can load up with more supplies. If the garbage is not removed, the townspeople will be in a very unhealthy environment. If the cells in your body cannot get rid of their waste products, they can actually be poisoned! On the next few pages, you will see how the urinary system removes waste materials from your blood so that the blood can transport nutrients again. The urinary system is shown in **Figure 11.**

— Kidneys
— Ureter
— Bladder
— Urethra

Figure 11 *The urinary system removes many of the waste products produced by the body.*

Flow-Through Filters

The **kidneys** are a pair of bean-shaped organs that constantly clean the blood. Your kidneys filter about 2,000 L of blood each day. Your body only holds 5.6 L of blood, so your blood cycles through the kidneys about 350 times a day!

Inside each kidney are more than 1 million microscopic filters called **nephrons,** shown below. Nephrons remove a variety of harmful substances from the body. Among the most important of these substances is urea, which contains nitrogen and is formed when cells use protein for energy.

Science CONNECTION

Why does your mouth get so dry when the rest of you is so hot and sweaty? Turn to page 72 to find out.

How the Kidneys Filter Blood

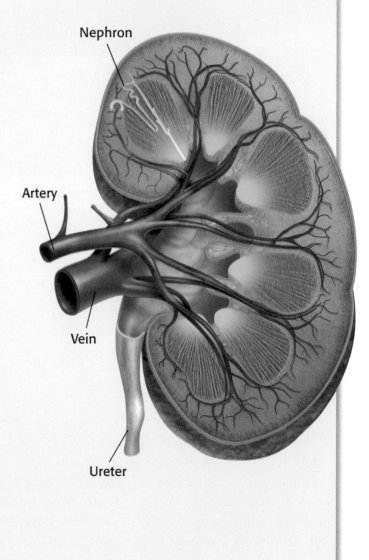

Nephron

Artery

Vein

Ureter

1 A large artery brings blood into each kidney.

2 Tiny blood vessels branch off the main artery and pass through part of each nephron.

3 Water and other small substances, such as glucose, salts, amino acids, and urea, are forced out of the blood vessels and into the nephrons.

4 As these substances flow through the nephrons, most of the water and some nutrients are moved back into blood vessels that wrap around the nephrons. A concentrated mixture of waste materials is left behind in the nephrons.

5 The cleaned blood, now with slightly less water and much less waste material, leaves each kidney in a large vein to recirculate in the body.

6 The yellow fluid that remains in the nephrons is called **urine.** Urine leaves each kidney through a slender tube called the *ureter* and flows into the **urinary bladder,** where it is stored.

7 Urine leaves the body through another tube called the *urethra. Urination* is the process of expelling urine from the body.

How Much Water?

Most adults need to consume about 2,500 mL of water per day. How many 8 oz glasses of drinking water would you need to get enough water? Hint: 1 oz = 29.6 mL

BRAIN FOOD

Fortunately, your kidneys are very efficient. In fact, if you lost one to disease, you could still survive.

Water In, Water Out

Our bodies take in water every day. If we did not excrete an equal amount of water, our bodies would swell up with all the excess. Losing water through sweat and urine is necessary to stay healthy. So how does the body keep the water levels in the proper balance? The balance of fluids is controlled by chemical messengers in the body called *hormones*.

Sweat and Thirst When you get hot, you lose more water in the form of sweat. The evaporation of water from your skin cools you down. As the water content of the blood drops, the salivary glands produce less saliva. This is one of the reasons you feel thirsty.

Antidiuretic Hormone When you get thirsty, other parts of your body react to the water shortage, too. A hormone called *antidiuretic* (AN tie DIE yoo RET ik) *hormone,* or *ADH,* is released. ADH signals the kidneys to take back water from the nephrons and return it to the bloodstream, thereby making less urine. When your blood is too watery, smaller amounts of ADH are released. The kidneys react by allowing more water to stay in the nephron and leave the body as urine.

Diuretics When you are thirsty, your tissues are asking for more water. Some beverages contain caffeine, which is a *diuretic* (DIE yoo RET ik). Diuretics cause the kidneys to make more urine, which decreases the amount of water in the blood. So instead of giving your body more water, caffeinated beverages cause additional water to be lost in urine.

APPLY

Beverage Ban

During football season, a football coach insists that all members of the team avoid caffeinated beverages. Many of the players are upset by the news. Pretend that you are the coach. Write a letter to the team explaining why it is better for them to drink water than drinks containing caffeine.

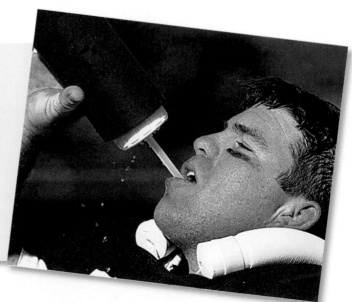

Urinary System Ailments

Since the urinary system regulates body fluids and removes wastes from the blood, any malfunction can become life-threatening. Some common urinary system disorders are described below.

Bacterial Infections Bacteria can get into the bladder and ureters through the urethra and cause painful infections. It is important to treat such an infection early because it could spread to the kidneys and lead to permanent damage to the nephrons.

Figure 12 *This kidney stone had to be removed from a patient's urinary system.*

Kidney Stones Sometimes salts and wastes collect inside the kidneys and form kidney stones, like the one in **Figure 12.** Kidney stones interfere with urine flow and cause pain. Most kidney stones pass naturally from the body, but sometimes a medical procedure is necessary. For example, shockwaves can be used to break the stones into pieces small enough to pass through the urethra.

Kidney Disease Damage to nephrons can prevent normal kidney functioning, leading to kidney disease. If the kidneys do not function properly, a kidney machine can be used to filter waste from the blood. As shown in **Figure 13,** blood is pumped from an artery in the forearm or wrist to a kidney machine, where it is filtered. The cleaned blood is then pumped back into a vein in the arm.

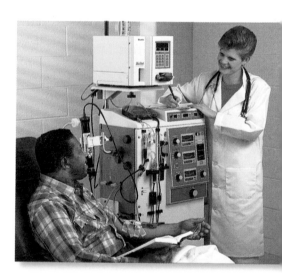

Figure 13 *The kidney machine will filter this man's blood before returning it to his body.*

SECTION REVIEW

1. Put the following statements about the urinary system in their proper order:

   a. Water is absorbed back into blood vessels.

   b. Urine leaves the kidney through the ureter.

   c. A large artery brings blood into the kidney.

   d. Water and other small substances, including glucose and urea, leave the blood vessels and enter the nephron.

2. What is the main function of the urinary system?

3. **Applying Concepts** Which contains more water, the blood going into the kidney or the blood leaving the kidney? Explain.

internet connect

SC*L*INKS
NSTA

TOPIC: The Urinary System, Urinary System Ailments
GO TO: www.scilinks.org
*sci*LINKS NUMBER: HSTL590, HSTL595

Skill Builder Lab

As the Stomach Churns

The stomach moves the food around while digestive juices—acids and enzymes—are added to begin protein digestion. Some meat tenderizers have plant enzymes that break down, or digest, proteins. You commonly can get two types of meat tenderizer at grocery stores. One type has an enzyme, called papain, from papaya. Another type has an enzyme, called bromelain, from pineapple. In this lab, you will test the effects of these two different types of meat tenderizers on beef stew meat.

MATERIALS

- 4 test tubes
- test-tube rack
- test-tube marker
- masking tape
- 25 mL graduated cylinder
- water
- eyedropper
- $\frac{1}{4}$ tsp measuring spoon
- hydrochloric acid (0.1 M)
- meat tenderizer containing bromelain
- meat tenderizer containing papain
- 1 cm cubes of beef stew meat (3)
- protective gloves

Ask a Question

1. Which meat tenderizer will work faster? Which will make the meat more tender? Will there be a color change in the meat or in the water? What might these changes, if any, indicate? Decide what you will look for as you plan your experiment.

Form a Hypothesis

2. Look at the list of ingredients on the labels of each of the meat tenderizers. Form a hypothesis about which tenderizer will make the beef more tender.
   Caution: Do not taste any of the materials in this lab.

Conduct an Experiment

3. Identify any variables and controls present in your experiment. Make a data table in your ScienceLog or on a computer to record your observations and results.

4. Label two of the test tubes with the name of the tenderizer being investigated. Label the third test tube "Control." What will this tube contain?

5 Pour 20 mL of water into each test tube.

6 With the eyedropper, add four drops of hydrochloric acid to each test tube. **Caution:** Hydrochloric acid can burn your skin. If any touches your skin, rinse the area with running water and tell your teacher immediately.

7 Using the measuring spoon, add $\frac{1}{4}$ tsp of a meat tenderizer to the tube with that label.

8 Add one piece of beef to each test tube.

9 Record your observations of each test tube immediately, after 5 minutes, after 15 minutes, after 30 minutes, and again after 24 hours.

Analyze the Results

10 Did you notice any differences in the beef in the three test tubes right away? At what time interval did you notice a significant difference in the appearance of the beef in the test tubes?

11 Did one meat tenderizer perform better than the other? Explain how you determined which tenderizer was more efficient.

Draw Conclusions

12 Was your hypothesis supported? Explain your answer.

13 Many stinging animals have venom composed of proteins. Explain how applying meat tenderizer to the wound helps relieve the pain of such a sting.

Chapter Highlights

Vocabulary

digestive system (*p. 54*)

esophagus (*p. 56*)

stomach (*p. 57*)

small intestine (*p. 58*)

pancreas (*p. 58*)

liver (*p. 59*)

gallbladder (*p. 59*)

large intestine (*p. 60*)

Section Notes

- Your digestive system is a group of organs that work together to digest food so that it can be used by the body.

- The breaking, crushing, and mashing of food is called mechanical digestion. Chemical digestion is the process in which large molecules are broken down to simpler molecules.

- Chewed food is pushed through the digestive tract by rhythmic contractions called peristalsis.

- The stomach mixes the food with enzymes and acid to break down nutrients. The mixture is called chyme.

- In the small intestine, pancreatic juice and bile are mixed with chyme.

- From the small intestine, nutrients enter the bloodstream and are circulated to the body's cells.

- The large intestine receives undigested material from the small intestine. As water is absorbed back into the body, this material becomes a solid mass called feces.

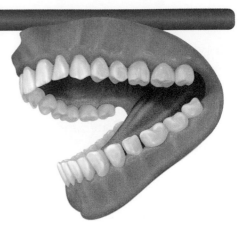

- Digestive system disorders include heartburn, constipation, diarrhea, colon cancer, and gastric ulcers.

Labs

Enzymes in Action (*p. 180*)

☑ Skills Check

Math Concepts

DRINK UP In the MathBreak on page 64, you determined how many glasses of water you need in order to drink 2,500 mL each day. First you must determine how many ounces are in 2,500 mL:

$$\frac{2{,}500 \text{ mL}}{1 \text{ day}} \times \frac{1 \text{ oz}}{29.6 \text{ mL}} = 84.5 \text{ oz of water}$$

You need to drink 84.5 oz of water each day.

Each glass contains 8 oz, so:

$$\frac{84.5 \text{ oz}}{1 \text{ day}} \times \frac{1 \text{ glass}}{8 \text{ oz}} = 10.6 \text{ glasses}$$

You need to drink 10.6 8-oz glasses of water each day.

Visual Understanding

KIDNEY FUNCTION Look at the illustration on page 63 to review how the kidneys filter the blood.

Vocabulary

urinary system (*p. 62*)

kidney (*p. 63*)

nephron (*p. 63*)

urine (*p. 63*)

urinary bladder (*p. 63*)

Section Notes

- Your skin, lungs, and urinary system are all involved in excretion.

- The urinary system cleans the blood and removes liquid waste as urine. The filtering structures in the kidneys are called nephrons.

- Most of the water and some nutrients that enter nephrons are moved back into the blood vessels.

- When urine leaves the kidneys, it passes into the urinary bladder through a tube called the ureter. The urinary bladder stores the urine until it can be eliminated.

- Urine travels from the urinary bladder to the outside through a tube called the urethra.

- Some disorders of the urinary system include bacterial infections, kidney stones, and kidney disease.

---

 internet connect

GO TO: go.hrw.com

Visit the **HRW** Web site for a variety of learning tools related to this chapter. Just type in the keyword:

KEYWORD: HSTBD3

SCi**LINKS**₅ₘ

N S T A

GO TO: www.scilinks.org

Visit the **National Science Teachers Association** on-line Web site for Internet resources related to this chapter. Just type in the *sci*LINKS number for more information about the topic:

TOPIC:	*sci*LINKS NUMBER:
The Digestive System	HSTL580
Problems in the Digestive System	HSTL585
The Urinary System	HSTL590
Urinary System Ailments	HSTL595
Tapeworms	HSTL600

Chapter Review

USING VOCABULARY

To complete the following sentences, choose the correct term from each pair of terms listed below:

1. Urine travels from each kidney to the urinary bladder through the __?__. (*urethra* or *ureter*)

2. The rhythmic contractions that occur in the digestive tract are called __?__. (*peristalsis* or *enzymes*)

3. The chemical digestion of carbohydrates begins in the __?__. (*stomach* or *mouth*)

4. Bile is made in the __?__ and stored in the __?__. (*liver* or *gallbladder*)

5. __?__ is the process of removing wastes and waste products from the body. This term is only used when substances are passed through a membrane before leaving the body. (*Digestion* or *Excretion*)

6. Indigestible material is formed into feces in the __?__. (*large intestine* or *small intestine*)

UNDERSTANDING CONCEPTS

Multiple Choice

7. The hormone that signals the kidneys to make less urine is
   a. urea.
   b. ADH.
   c. cellulase.
   d. ATP.

8. Which of the following aids digestion by producing bile?
   a. stomach
   b. pancreas
   c. gallbladder
   d. liver

9. The part of the kidney that filters the blood is the
   a. artery.
   b. ureter.
   c. nephron.
   d. urethra.

10. The fingerlike projections lining the small intestine are called
    a. emulsifiers.
    b. fats.
    c. amino acids.
    d. villi.

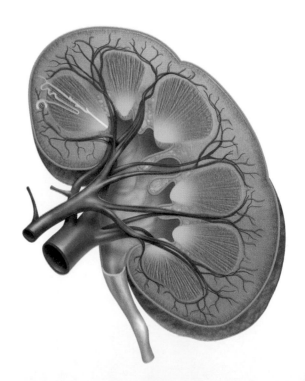

11. Which of the following is not part of the digestive tract?
    a. mouth
    b. pancreas
    c. stomach
    d. rectum

12. The soupy mixture of food, enzymes, and acids in the stomach is called
    a. chyme.
    b. villi.
    c. urea.
    d. vitamins.

Short Answer

13. Give two reasons why it is important that the pancreas releases bicarbonate into the small intestine.

14. How does the structure of the small intestine improve its nutrient absorption?

15. What is a diuretic?

Concept Mapping

16. Use the following terms to create a concept map: teeth, stomach, digestion, bile, saliva, mechanical digestion, gallbladder, chemical digestion.

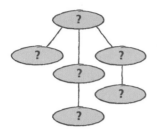

CRITICAL THINKING AND PROBLEM SOLVING

Write one or two sentences to answer the following questions:

17. How would digestion be affected if the liver were damaged?

18. Think about what happens when you put a piece of carbohydrate-dense food, such as bread, potato, or a cracker, in your mouth. If you let a small piece sit near the tip of your tongue, it might begin to taste sweet. What digestive process would explain this change?

MATH IN SCIENCE

19. Mr. Jones has lost all of his molars and two of his premolars. How many teeth does Mr. Jones have?

20. During a one-day water-balance study, a woman drank 1,500 mL of water. The food that she ate contained 750 mL of water, and 250 mL of water was produced internally during normal body processes. She lost 900 mL of water by sweating, 1,500 mL in urine, and 100 mL in feces. Overall, how much water did she gain or lose during the day?

INTERPRETING GRAPHICS

The bar graph below shows how long the average meal spends in each portion of your digestive tract. Use this graph to answer the questions below.

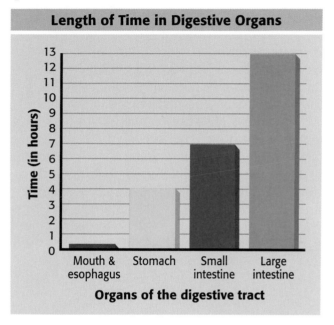

21. Where does the food spend the longest amount of time?

22. On average, how much longer does food stay in the small intestine than in the stomach?

23. Which organ mixes food with special substances to make chyme? Approximately how long does food remain in this organ?

24. Bile breaks up large fat droplets into very small droplets. How long is the food in your body before it comes into contact with bile?

Reading Check-up

Take a minute to review your answers to the Pre-Reading Questions found at the bottom of page 52. Have your answers changed? If necessary, revise your answers based on what you have learned since you began this chapter.

The Digestive and Urinary Systems   **71**

Quench Your Thirst!

Have you ever been really thirsty after a hard workout? Playing sports, riding a bike, and doing other physical activities can make you thirsty—but why? The first reason is sweat. When you are physically active, you lose a lot of water by sweating. This keeps your body from overheating. But what is going on in your body to make your mouth feel so dry? And which is better to quench your thirst, water or a sports drink?

▲ *Activities that make you hot and sweaty make you want to take a drink, but why?*

Thirsty Chemistry

When you lose water, your blood becomes more concentrated. Think about how you make a powdered drink, such as lemonade. If you use the same amount of powder in 1 L of water as you do in 2 L, the drinks will taste different. The lemonade made with 1 L of water will be stronger because it is more concentrated.

Losing water to sweat increases the concentration of sodium and potassium in your blood. The kidneys force the extra potassium out of the blood vessels and into nephrons. From the nephrons, the potassium is eliminated from the body in urine. Nerve cells in your brain react to the high concentration of sodium by sending out two important messages. One message tells the pituitary gland to release antidiuretic hormone. This hormone signals the kidneys to return water to the bloodstream. The second message signals the salivary glands to produce less saliva. With less saliva, your mouth becomes dry, and then you know it's time to get that drink!

With Flavor or Without?

But which is better to drink—water or a sports drink? If you have been exercising, you might think a sports drink is probably better. But studies by Kathy Grunewald, a professor at Kansas State University, indicate that sports drinks may not be necessary unless you have exercised very hard or for more than an hour and a half.

When you drink fluids, you lower the concentration of all the minerals in your blood by adding more water. This is like adding water to the 1 L of strong lemonade. The body also needs to replace the potassium that was lost from the blood. A sports drink can help replace this potassium. But if you drink water, the kidneys will eventually return the potassium concentration to normal. The most important reason for drinking fluids after physical exercise is to get water to your tissues. So whatever physical activity you choose, drink up!

Going Further

▶ If you want to investigate how much you need to drink, weigh yourself before and after your next strenuous activity. Every kilogram that you lose represents about 1 L of water. You should make sure that you drink at least as much water as you lose.

Health

A Voiceless Companion

If you decided to eat the last piece of pizza in the refrigerator, someone just might ask you for a bite, right? But what if you found out that you had a constant mealtime companion who didn't want just a bite but wanted it all? And what if that companion never asked for your permission?

How to Be a Host

This constant mealtime companion might be a tapeworm. Tapeworms are invertebrate flatworms. These flatworms are parasites. A parasite is an organism that obtains its food by living in or on another living organism. The organism in which a parasite makes its home is called a host. People, cows, pigs, fish, cats, dogs, and many insects are the perfect hosts for tapeworms. Without a host, tapeworms can't survive.

Food broken down in the stomach continues to be broken down in the small intestine. Since a tapeworm doesn't have a digestive tract of its own, it borrows its host's. By attaching itself to the inside of its host's small intestine with clamps and suckers, a tapeworm can eat as much as it likes.

Although tapeworms aren't much thicker than a ribbon, they can grow to more than 6 m in length! They do this by adding one postage-stamp-sized segment at a time. Each segment has both male and female reproductive organs and can be filled with thousands of eggs.

Saying "Goodbye" and Avoiding "Hello"

When an egg-filled segment breaks off, it passes through the rest of the host's digestive tract and ends up in the feces. If another animal eats or drinks something contaminated with these feces, the eggs grow into worms in that animal's intestines. The eggs can then spread to muscle tissues (called meat in animals used for food). If humans eat this meat but don't cook it thoroughly enough to kill tapeworm larvae, the cycle begins all over again.

Getting rid of a tapeworm requires removing the head, or scolex. If the scolex is left behind, it simply produces new segments, and the tapeworm regrows itself. Sometimes humans don't realize they have a tapeworm, even though they suffer from symptoms such as weight loss and nausea. And occasionally there are no symptoms at all.

The best way to avoid these parasites is to avoid eating undercooked beef, pork, and fish. If you do this, you won't have any uninvited guests at your next meal!

▲ *What is 10 m long, looks like it's made of postage stamps, and eats your dinner after you do?*

Think About It

▶ Doctors prescribe certain medications to get rid of tapeworms. Research the different ways people got rid of tapeworms before modern medicines were available.

CHAPTER 4

Communication and Control

Pre-Reading Questions

1. What are your senses? How do senses help us survive?

2. Why does your heart beat faster when something frightens you?

3. How do eyeglasses and contact lenses help some people see better?

OUTTA SIGHT!

This may look like a flower garden or an oceanic reef. But it's really something much closer to home. It's the human tongue (magnified thousands of times, of course). You know these round bumps as *taste buds*. You use taste and other senses to gather information about your surroundings. This information helps your body respond to its environment. In this chapter, you will find out how the human body senses the world and controls its own functions.

ACT FAST!

If you want to catch an object, your brain sends a message to your arm's muscles. In this exercise, you will see how long that takes.

Procedure

1. Sit in a **chair** with one arm in a "handshake" position. Your partner should stand facing you, holding a **meterstick** vertically. The stick should be positioned to fall between your thumb and fingers.

2. Tell your partner to let go of the meterstick without warning. Catch the stick between your thumb and fingers. Your partner should catch the meterstick if it tips over.

3. Record the number of centimeters the stick dropped before you caught it. That distance represents your reaction time.

4. Repeat steps 1–3 three times. Calculate the average distance.

5. Repeat steps 1–4 with your other hand.

6. Trade places with your partner, and repeat steps 1–5.

Analysis

7. Compare the reaction times of your own hands. Why might one hand react faster?

8. Compare your results with your partner's. Why might one person react faster than another?

Terms to Learn

central nervous system
peripheral nervous system
neuron
impulse
receptor
nerve
brain
reflex

What You'll Do

◆ Explain how neurons in the nervous system work together.
◆ Compare the central nervous system with the peripheral nervous system.
◆ Describe the major functions of the brain and the spinal cord.

The Nervous System

What do the following events have in common? You hear a knock at the door, you write a book report, you feel your heart pounding after a run, you work a math problem, you are startled by a loud noise, and you enjoy eating a sweet mango. These events are all activities of your nervous system. The nervous system gathers and interprets information about the body's internal and external environments and responds to that information. The nervous system keeps your organs working properly and allows you to speak, smell, taste, hear, see, move, think, and experience emotions.

Two Systems Within a System

Your nervous system controls and coordinates many things that happen in your body. It acts as a central command post, collecting and processing information and making sure appropriate information gets sent to all parts of the body. These tasks are accomplished by two subdivisions of the nervous system, the *central nervous system* and the *peripheral nervous system*.

The **central nervous system** (CNS) includes your brain and spinal cord. It processes all incoming and outgoing messages. The **peripheral nervous system** (PNS) consists of communication pathways, or *nerves*, that connect all areas of your body to your CNS. **Figure 1** shows the major divisions of the nervous system.

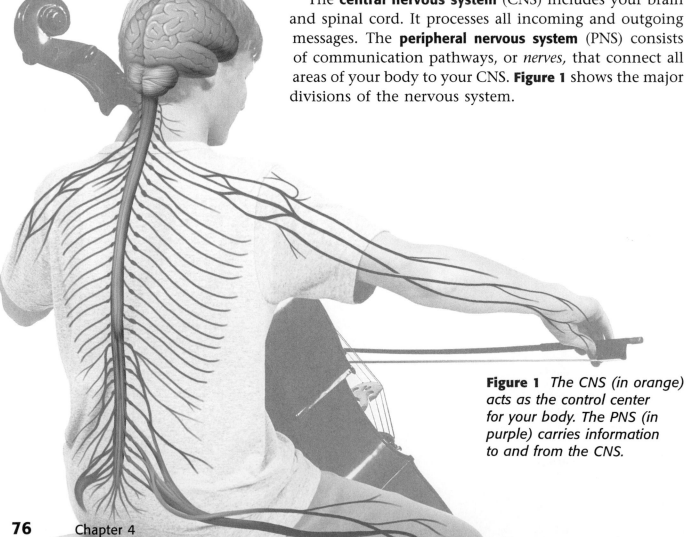

Figure 1 *The CNS (in orange) acts as the control center for your body. The PNS (in purple) carries information to and from the CNS.*

The Peripheral Nervous System

How long does it take for a light to come on when you flip a light switch? The light seems to come on immediately. In a similar way, specialized cells called **neurons** transfer messages throughout your body in the form of fast-moving electrical energy. A typical neuron is shown in **Figure 2.** The electrical messages that pass along the neurons are called **impulses.** Impulses may travel as fast as 150 m/s or as slow as 1 m/s.

Neuron Structure A neuron consists of a cell body, dendrites, and axons. The enlarged region called the cell body contains a nucleus and cell organelles. Look again at Figure 2. The neuron generally receives information from other cells through short, branched extensions called *dendrites*. A neuron may have many dendrites, allowing it to receive impulses from thousands of other cells.

From the cell body, information is transmitted to other cells by a fiber called an *axon*. Axons can be very short or quite long. You have some really long axons that extend almost 1 m from your lower back to your toes. The end of an axon often has branches that allow information to pass to yet more cells. The tip of each branch is called an *axon terminal*.

÷ 5 ÷ Ω ≤ ∞ +Ω √ 9 ∞ ≤ Σ 2
+

MATH BREAK

Time to Travel

To calculate how long it takes for an impulse to travel a certain distance, you can use the following equation:

$$\text{Time} = \frac{\text{distance}}{\text{speed}}$$

If an impulse travels 100 m/s, about how long would it take for an impulse to travel 10 m?

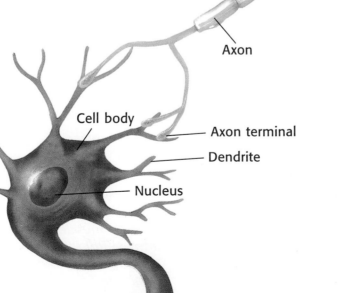
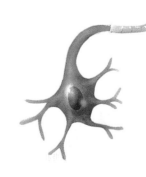

Figure 2 *Neurons are special cells that transfer electrical messages throughout the body.*

Direction of impulse

Axon

Cell body

Axon terminal

Dendrite

Nucleus

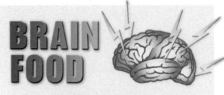
Information Collection   Special neurons called *sensory neurons* gather information about what is happening in and around your body and send this information on to the central nervous system for processing. Sensory neurons have specialized dendrites called **receptors** that detect changes inside and outside the body. For example, receptors in your eyes detect the light around you. Receptors in your stomach let your brain know when your stomach is full or empty.

Delivering Orders   Neurons that send impulses from the brain and spinal cord to other systems are called *motor neurons*. Motor means "to move"; when muscles get impulses from motor neurons, they respond by contracting. For example, motor neurons cause the muscles around your eyes to move when the sensory neurons in your eyes detect bright light. This movement makes you squint, which reduces the amount of light entering the eye. Motor neurons also send messages to your glands, such as sweat glands. These messages tell the sweat glands to release sweat.

Just a Bundle of Axons

The central nervous system is connected to the rest of your body by nerves. **Nerves** are axons bundled together with blood vessels and connective tissue. Nerves extend throughout your body. Most nerves contain the axons of both sensory and motor neurons. **Figure 3** shows the structure of a nerve. The axon in this nerve transmits information from the spinal cord to muscle fibers.

Spinal cord

Nerve

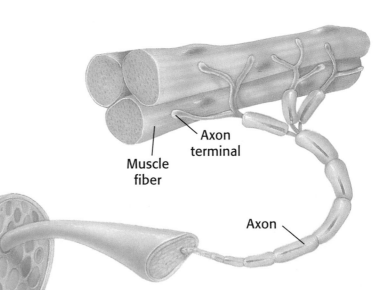

Axon terminal

Muscle fiber

Axon

Figure 3 *In order for a muscle to contract, a message must travel from the spinal cord to the muscle. The message travels along the axon of a motor neuron inside the nerve.*

The Central Nervous System

The central nervous system works closely with the peripheral nervous system. It receives information from the sensory neurons and responds by sending messages to various parts of the body via motor neurons.

Mission Control The **brain,** part of your central nervous system, is the nervous system's largest organ. It has hundreds of different jobs. Many of the processes that the brain controls happen automatically and are referred to as *involuntary.* For example, you couldn't stop digesting the food you have eaten even if you tried. Other activities controlled by your brain are *voluntary.* When you want to move your arm, your brain sends signals along motor neurons to muscles in your arm. This causes the muscles to contract and your arm to move. The brain has three connected parts—the cerebrum, the cerebellum, and the medulla. Each part has its own functions.

Your Thinking Cap The largest part of your brain is called the *cerebrum.* Its shape resembles a mushroom cap over a stalk. This dome-shaped area is where you think and where most memories are stored. It controls voluntary movements and allows you to detect touch, light, sound, odors, taste, pain, heat, and cold.

The cerebrum has two halves called *hemispheres.* The left hemisphere directs the right side of the body, and the right hemisphere directs the left side of the body. This is because axons cross over to the opposite side of the body in the spinal cord. **Figure 4** gives a general model of the activities that each hemisphere controls. However, most brain activity involves both hemispheres.

BRAIN FOOD

The organism with the largest brain is the sperm whale. Its brain is six times the size of a human brain!

▼ The **right hemisphere** primarily controls activities that involve imagination, appreciation, and creativity.

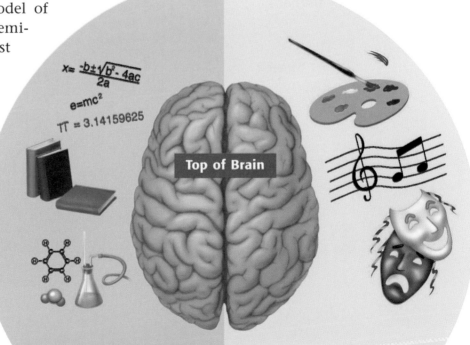

**Figure 4
The Cerebral Hemispheres**

$$x = \frac{-b \pm \sqrt{b^2 - 4ac}}{2a}$$

$e = mc^2$

$\pi = 3.14159625$

Top of Brain

The **left hemisphere** ▶ primarily controls activities such as speaking, reading, writing, and solving problems.

The Balancing Act The second largest part of your brain is the *cerebellum* (SER uh BEL uhm). It lies underneath the back of your cerebrum and receives sensory impulses from skeletal muscles and joints. This allows the brain to keep track of your body's position. For example, if you begin to lose your balance, like the girl in **Figure 5,** the cerebellum sends impulses to different skeletal muscles to make them contract, keeping you upright.

The Mighty Medulla The part of your brain that connects to your spinal cord is called the *medulla* (mi DOOL uh). The medulla is only about 3 cm long, but you couldn't live without it. The medulla controls your blood pressure, heart rate, involuntary breathing, and some other involuntary activities.

Your medulla constantly receives sensory impulses from receptors in your blood vessels. It uses this information to regulate your blood pressure. If your blood pressure gets too low, the medulla sends out impulses that tell blood vessels to tighten up to increase the blood pressure. The medulla also sends impulses to the heart to make it beat faster or slower as necessary. **Figure 6** shows the location of each part of the brain and some of the functions associated with each part.

Figure 5 *Your cerebellum causes skeletal muscles to make adjustments in order to keep you upright.*

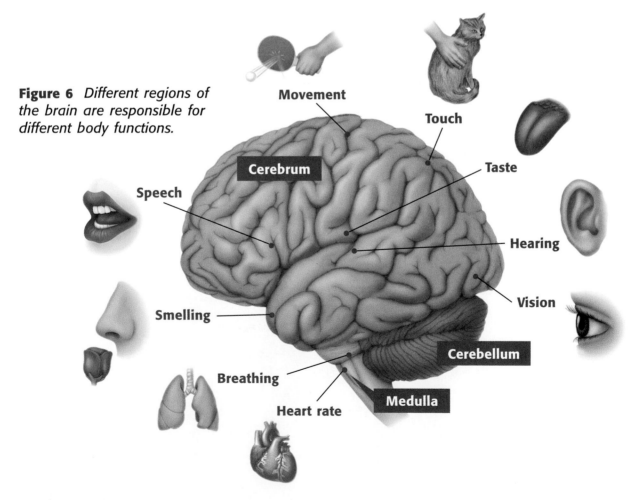

Figure 6 *Different regions of the brain are responsible for different body functions.*

Movement

Touch

Cerebrum

Taste

Speech

Hearing

Smelling

Vision

Cerebellum

Breathing

Medulla

Heart rate

The Spinal Cord

Your spinal cord, part of the central nervous system, is about as big around as your thumb. It contains neurons and bundles of axons that pass impulses to and from the brain. As shown in **Figure 7,** the spinal cord is surrounded by protective bones called *vertebrae* (VUHR tuh BRAY).

The nerve fibers in your spinal cord enable your brain to communicate with your peripheral nervous system. Sensory neurons in your skin and muscles send impulses along their axons to your spinal cord. The spinal cord then conducts impulses to your brain, where they can be interpreted as pain, heat, cold, or other sensations. Impulses moving from the brain down the cord are relayed to motor neurons, which carry the impulses along their axons to muscles and glands all over your body.

Spinal Cord Injury If the spinal cord is injured, any sensory information coming into it below where the damage occurred may be unable to travel to the brain. Likewise, any motor commands the brain sends to an area below the injury may not get through to the peripheral nerves. Thousands of people each year are paralyzed by spinal cord injuries. Many of these injuries occur in automobile accidents. Among young people, spinal cord injuries are often sports related.

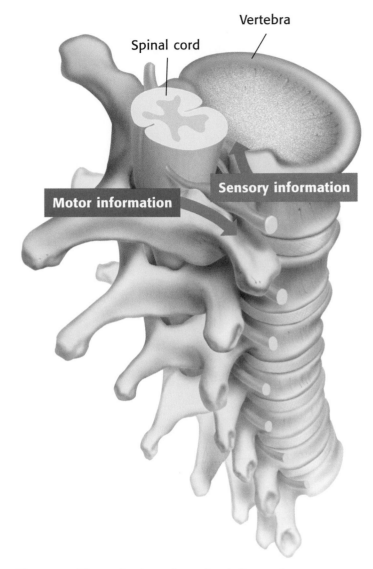

Vertebra

Spinal cord

Sensory information

Motor information

Figure 7 *The spinal cord carries information to and from the brain. It is protected by vertebrae.*

Self-Check

1. What part of the brain do you use to do your math homework?

2. What part of the brain helps a gymnast maintain balance on the balance beam?

3. What is the function of the vertebrae?

(See page 212 to check your answers.)

Ouch! That Hurt!

Have you ever stepped on something sharp? You probably pulled your foot up without thinking. This quick, involuntary action is called a **reflex.** Reflexes help protect your body from damage.

When you step on a sharp object, the message "pain" travels to your spinal cord, and a message to move your foot travels back to the muscles in your leg. The muscles in your leg respond before the information ever reaches the brain. By the time your brain finds out what happened, your foot has already moved. If you had to wait for your brain to get the message, your foot might be seriously injured! The man in **Figure 8** lifted his foot before he realized he had stepped on a toy.

Figure 8 *When pain impulses from your foot reach your spine, a message is sent immediately to your leg muscles to lift your foot.*

Knee Jerks

1. Sit on the edge of a desk or table so your feet don't touch the floor.

2. While your leg is completely relaxed, have a classmate *gently* tap on your knee slightly below the kneecap with the edge of his or her hand. How did your leg respond? Did you have any control over what happened? Explain.

3. Describe the pathway taken by the impulse that started with the tap on the knee.

SECTION REVIEW

1. Make a labeled diagram that shows the path of an electrical message from one neuron to another neuron.

2. Explain how the peripheral nervous system connects with the central nervous system.

3. If a spider is crawling up your left arm, which cerebral hemisphere controls the movement that you will use to knock it off?

4. List the three major parts of the brain, and describe their functions.

5. **Applying Concepts** Describe a time when you experienced a reflex.

Terms to Learn

retina   iris
rods   lens
cones   cochlea

What You'll Do

◆ List four sensations that are detected by receptors in the skin.
◆ Describe how light relates to vision.
◆ Explain the functions of photoreceptors, taste buds, and olfactory cells.

Responding to the Environment

How do you know when someone taps you on the shoulder or calls your name? How do you feel the touch or hear the sound? Impulses from sensory receptors in your shoulder and in your ears travel to your brain, sending information about your external environment. Your brain depends on this information to make decisions that affect your survival.

Come to Your Senses

Information about your surroundings and the conditions in your body is detected by sensory receptors. This information is converted to electrical signals and sent to your brain for interpretation. Once the signals reach your brain, you become aware of them. This awareness is called a *sensation*. It is in your brain that you have thoughts, feelings, and memories about sensations.

There are many different kinds of sensory receptors in your body. For example, receptors in your eyes detect light. Receptors in your ears detect vibrations called sound waves. The taste buds on your tongue have receptors that detect chemicals in the foods you eat. You have special receptors in your nose that detect tiny particles in the air. Your skin has a variety of receptors as well. Look at **Figure 9** to see some of the different kinds of receptors in the skin.

Figure 9 *This diagram shows some of the receptors in your skin and what they detect.*

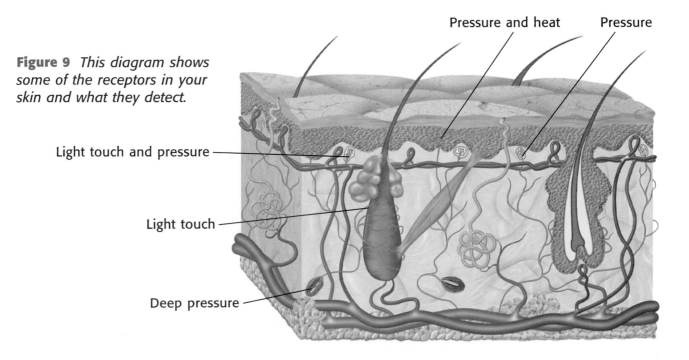

Pressure and heat   Pressure

Light touch and pressure

Light touch

Deep pressure

Communication and Control **83**

Something in My Eye

As you read this sentence, you are using one of your most important senses—*vision*. Vision is your awareness of light energy. Your eyes have special receptors that detect visible light, a portion of the sun's energy that reaches the Earth.

An Eyeful The eye is a complex sensory organ. Examine the eye in **Figure 10.** The outer surface of the eye is covered by the cornea, a transparent membrane that protects the eye but allows light to enter. Visible light that is reflected by objects around you enters through an opening at the front of your eye called the *pupil*. Light is detected by cells at the back of your eye in a light-sensitive layer called the **retina.**

The retina is packed with special neurons called *photoreceptors* (*photo* means "light") that convert light into electrical impulses. There are two types of photoreceptors in the retina—rods and cones. **Rods** can detect very dim light. They are important for night vision. Impulses from rods are perceived in tones of gray. In bright light, the **cones** give you a very colorful view of the world.

Light energy produces changes in photoreceptors that trigger nerve impulses. These impulses travel along axons, leaving the back of each eye through an *optic nerve.*

BRAIN FOOD

Carrots and other foods rich in vitamin A can improve your night vision. Vitamin A is important in maintaining proper functioning of the rods in your retina.

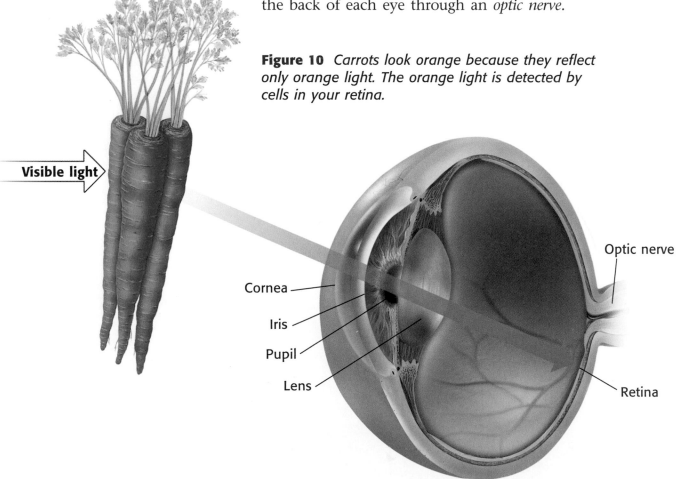

Figure 10 *Carrots look orange because they reflect only orange light. The orange light is detected by cells in your retina.*

Visible light

Cornea

Iris

Pupil

Lens

Optic nerve

Retina

Seeing the Light Light rays enter the eye through the *pupil*. Your pupil looks like a black dot in the center of your eye, but it is actually an opening. It is surrounded by the **iris,** the part of the eye that gives the eye color. A ring of muscle fibers causes the iris to open and close, making the pupil change size. This regulates the amount of light that passes to the retina. In bright light, your pupil is small, and in dim light, your pupil is large.

Hocus Focus Light travels in straight lines until it passes through the cornea and the *lens*. A **lens** is a piece of curved material behind the pupil that allows light to pass through but changes its direction. The lens focuses the light entering the eye on the retina. The lens of an eye changes shape to adjust focus. When you look at objects close to the eye, the lens becomes more curved. When you look at objects far away, the lens gets flatter.

In some eyes, the lens focuses the light just in front of the retina (resulting in nearsightedness) or just behind the retina (resulting in farsightedness). Glasses or contact lenses can usually correct these vision problems. Focus on **Figure 11** to see how corrective lenses work.

What are some other uses of lenses? Turn to Light on Lenses on page 98 to find out.

Figure 11 *A concave lens bends light rays outward to correct nearsightedness. A convex lens bends light rays inward to correct farsightedness.*

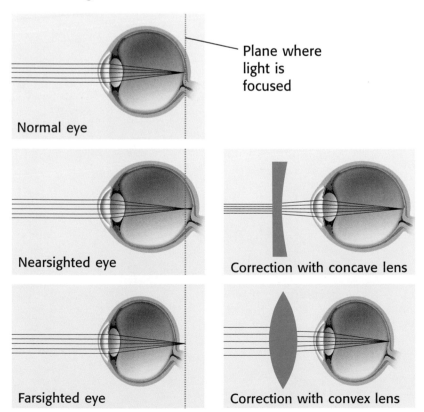

Plane where light is focused

Normal eye

Nearsighted eye

Correction with concave lens

Farsighted eye

Correction with convex lens

Quick Lab

Where's the Dot?

1. Hold this book at arm's length, and close your right eye. Focus your left eye on the solid dot below.

2. Slowly move the book toward your face. Stay focused on the solid dot.
3. What happens to the white dot?
4. Do some research on the optic nerve to find out why this happens.

Try at Home

Did You "Ear" That?

When a guitar string is plucked, what enables you to hear the sound? A sound begins when an object, such as the guitar string, begins to vibrate. The vibrations push on surrounding air particles. These air particles push on other air particles, transferring energy in waves away from the source. Hearing is the sensation experienced in response to these sound waves.

Journey of a Sound Wave Your ears are organs specialized for hearing. Each ear has an outer, middle, and inner part. The parts of the ear are shown in **Figure 12.** When sound waves reach the outer ear, they are funneled into the middle ear, where they cause the eardrum to vibrate. The vibrating eardrum makes tiny ear bones vibrate. One of the tiny bones vibrates against the **cochlea** (KAHK lee uh), a tiny snail-shaped organ of the inner ear. Inside the cochlea, the vibrations create waves that are similar to the waves you can make by tapping on a glass of water. Neurons in the cochlea convert these waves to electrical impulses and send them to the area of the brain that interprets sound.

Figure 12 *A sound wave travels into the outer ear. It is converted to bone vibrations in the middle ear, then to liquid vibrations in the inner ear, and finally to nerve impulses.*

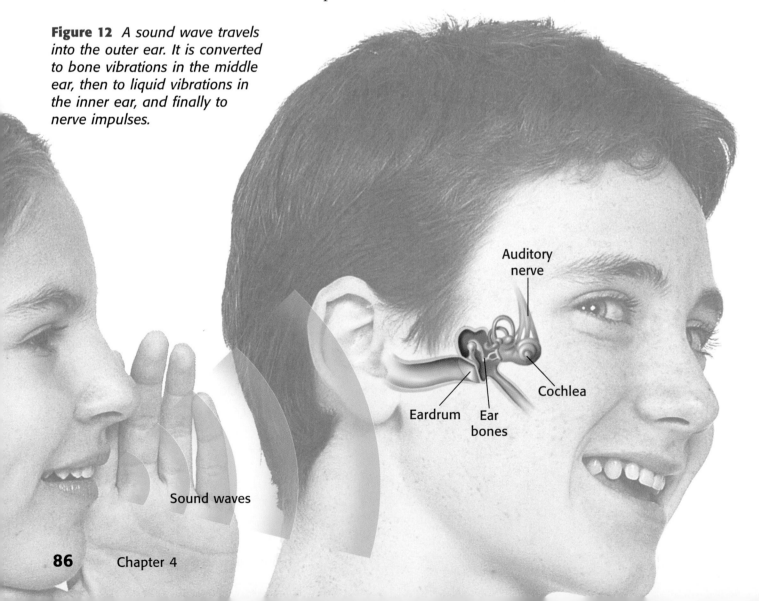

Auditory nerve

Cochlea

Eardrum   Ear bones

Sound waves

Does This Suit Your Taste?

When you put food in your mouth, your sense of what the food tastes like comes mostly from your tongue. Taste is the sensation you feel when the brain is made aware of certain dissolved chemicals in your mouth. The receptors for taste are clustered in the *taste buds*. The tongue is covered with tiny bumps called *papillae* (puh PIL ee), and the taste buds are embedded in the sides of these bumps. As shown in **Figure 13,** there are four types of taste buds. Each type responds to one of the four basic tastes: sweet, sour, salty, and bitter.

Your Nose Knows

Have you ever noticed that when you have a congested nose you can't taste food very well? Try eating a piece of peppermint while holding your nose. The mint taste is not very intense until you inhale through your nose. That's because smell and taste are closely related. The brain combines information from your taste buds and nose to give you a sense of flavor. The receptors for smell are located on *olfactory cells* in the upper part of your nasal cavity. They react to chemicals that are inhaled and dissolved in the moist lining of the nasal cavity. The woman in **Figure 14** is using her sense of smell to test the effectiveness of underarm deodorants.

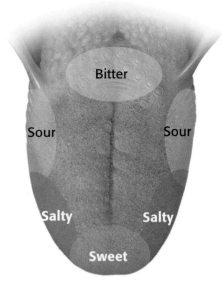

Figure 13 *Taste buds in different parts of the tongue respond to different types of chemicals.*

Figure 14 *This woman's nose is detecting chemicals in the sweat and in the deodorants used by this man. Her brain will generate opinions about the smells that she will then record in her report.*

SECTION REVIEW

1. List three sensations that receptors in the skin can detect.

2. Explain why you would have trouble seeing bright colors at a candlelit dinner.

3. How is your sense of taste similar to your sense of smell?

4. **Applying Concepts** If you can focus on objects close to you but things become blurry when they are far away, would a concave or convex lens correct your vision?

🖵 internet**connect**

SC/LINKS.
NSTA

TOPIC: The Senses, The Eye
GO TO: www.scilinks.org
sciLINKS NUMBER: HSTL610, HSTL615

SECTION 3

READING WARM-UP

Terms to Learn

endocrine system
gland
hormone
feedback control

What You'll Do

- Explain the function of the endocrine system.
- List the glands of the endocrine system and describe some of their functions.
- Describe how feedback controls stop and start hormone release.

The Endocrine System

You already know that the job of the nervous system is to communicate with all the other body systems. Its main role is to respond to stimuli. But it is not the only system that has this role. Your **endocrine system** is involved with the control of slower, long-term processes, such as fluid balance, growth, and sexual development. Instead of electrical messages, the endocrine system sends messages via chemicals.

Chemical Messengers

The endocrine system controls body functions with the use of chemicals that are released from endocrine glands. A **gland** is a group of cells that makes special chemicals for your body. Chemicals that are produced by the endocrine glands are called **hormones.** The chemicals made by endocrine glands are released into the bloodstream and carried to other places in the body. Because hormones act as chemical messengers, an endocrine gland near your brain can control the actions of an organ located somewhere else in your body.

Glands at Work Endocrine glands often affect many organs at one time. For example, your adrenal glands prepare your organs to deal with stress. They make the hormone *epinephrine* (ep ih NEF rihn), also known as *adrenalin.* Epinephrine speeds up your heartbeat and breathing rate to prepare your body either to run from danger or to fight for survival. This hormone effect is often referred to as the fight-or-flight response. You may have noticed these effects when you were frightened or angry.

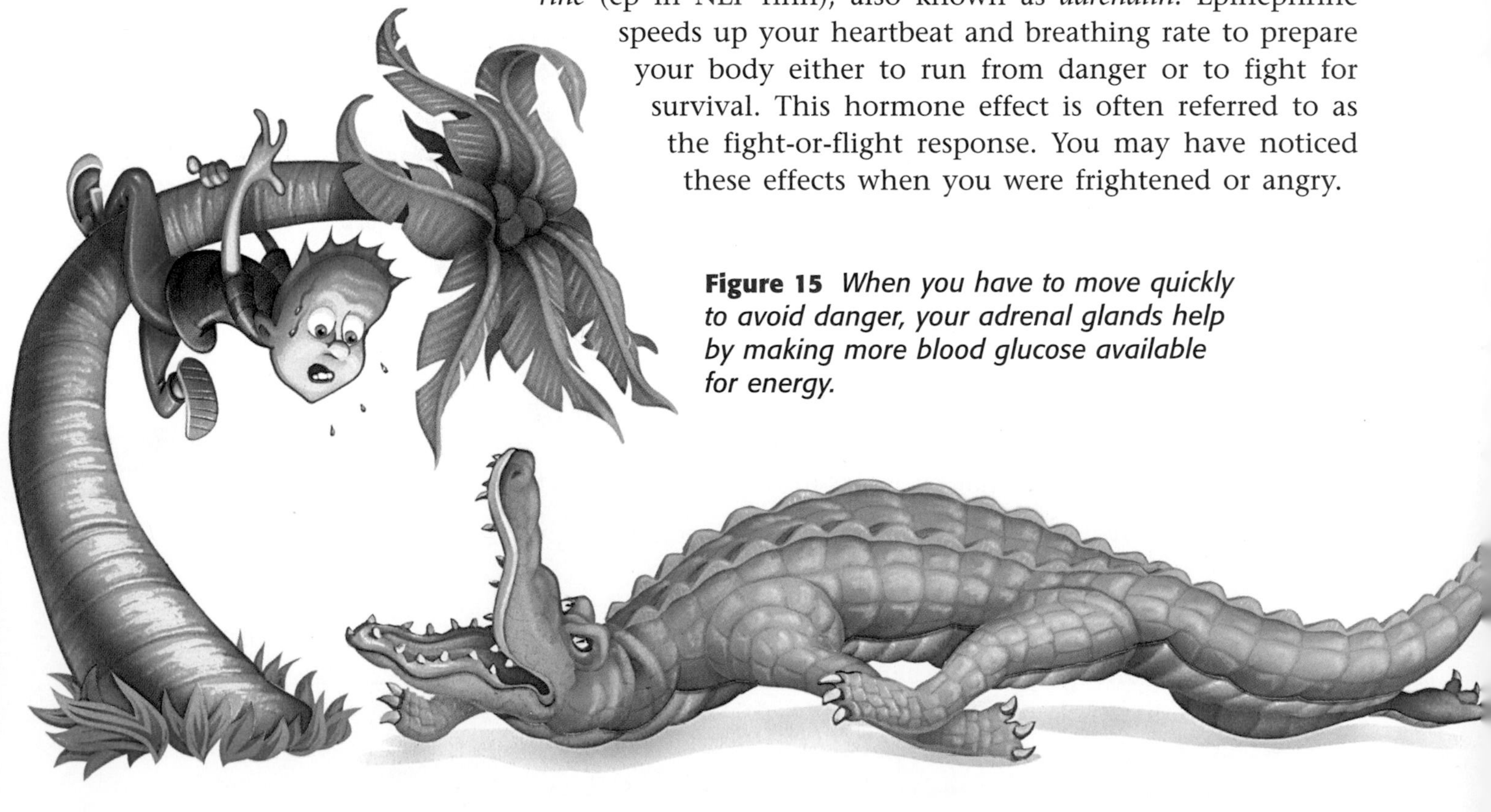

Figure 15 *When you have to move quickly to avoid danger, your adrenal glands help by making more blood glucose available for energy.*

Fight or Flight?

Maria was working late at the library. She was worried about walking home alone. As she started home, she noticed a shadowy figure walking quickly behind her. The figure was gaining on her! She could feel her heart pounding in her chest. She began to run, and then a familiar voice called out her name. It was her father. He had walked to the library to check on her. What a relief!

Maria had a fight-or-flight response. Write a paragraph describing a time when you had a fight-or-flight experience. Include in your story the following terms: *hormones, fight-or-flight,* and *epinephrine.*

Your body has several other endocrine glands, some with many different functions. For example, your pituitary gland stimulates skeletal growth, helps the thyroid function properly, regulates the amount of water in the blood, and stimulates the birth process in pregnant women. The names and some of the functions of this and other endocrine glands are summarized in **Figure 16.**

Figure 16 *Your endocrine glands produce chemicals called hormones that control many of your body functions.*

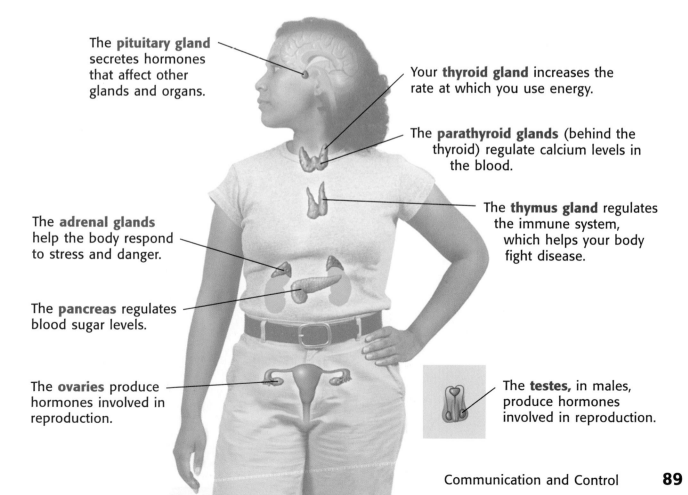

The **pituitary gland** secretes hormones that affect other glands and organs.

Your **thyroid gland** increases the rate at which you use energy.

The **parathyroid glands** (behind the thyroid) regulate calcium levels in the blood.

The **thymus gland** regulates the immune system, which helps your body fight disease.

The **adrenal glands** help the body respond to stress and danger.

The **pancreas** regulates blood sugar levels.

The **ovaries** produce hormones involved in reproduction.

The **testes,** in males, produce hormones involved in reproduction.

Controlling the Controls

How do endocrine glands know when to start and stop hormone release? They know because your body has special systems called **feedback controls** that turn endocrine glands on and off. Feedback controls work something like a thermostat on an air conditioner. Once a room reaches the required temperature, the thermostat sends a message to the air conditioner to stop sending in cold air. Much in the same way, a feedback control sends a message to an endocrine gland to stop sending in a particular hormone. **Figure 17** traces the steps of a feedback control that regulates blood sugar.

Figure 17 *In this feedback-control system, the pancreas produces hormones that help your body maintain the correct blood sugar level.*

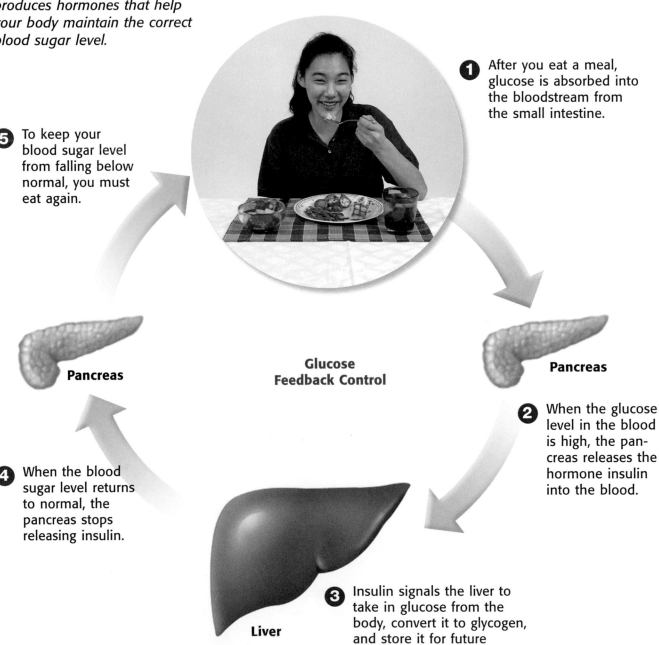

Glucose Feedback Control

❶ After you eat a meal, glucose is absorbed into the bloodstream from the small intestine.

❷ When the glucose level in the blood is high, the pancreas releases the hormone insulin into the blood.

❸ Insulin signals the liver to take in glucose from the body, convert it to glycogen, and store it for future energy needs.

❹ When the blood sugar level returns to normal, the pancreas stops releasing insulin.

❺ To keep your blood sugar level from falling below normal, you must eat again.

Pancreas

Pancreas

Liver

Hormone Imbalances

Insulin is a hormone made by the pancreas. When the blood sugar level rises after a person has eaten something, insulin triggers the cells to take in glucose and sends a message to the liver to store glucose. A person whose pancreas cannot make enough insulin has a condition called *diabetes mellitus*. A person with diabetes mellitus may need daily injections of insulin to keep his or her blood glucose levels within safe limits. Some patients, like the woman in **Figure 18,** receive their insulin automatically from a small machine they wear next to their body.

Growth Hormone Sometimes a child may have a pituitary gland that doesn't make enough growth hormone. This causes the child's growth to be stunted. Fortunately, if this problem is detected soon enough, a doctor can prescribe hormone replacement medication and monitor the child's growth. If the pituitary makes too much growth hormone at an early age, the person becomes much taller than expected.

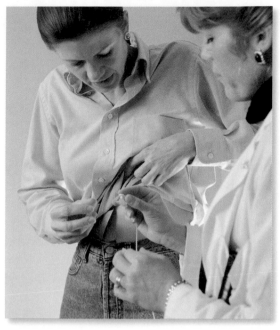

Figure 18 *This young woman has diabetes and must have daily injections of the hormone insulin.*

Thyroxine When a person doesn't get enough iodine in the diet, the thyroid gland cannot make enough of the hormone *thyroxine*. This causes the thyroid to swell up and form a mass called a *goiter*. Because thyroxine increases metabolism, this person's cells are less active than normal, causing fatigue, weight gain, and other problems.

SECTION REVIEW

1. What is the function of the endocrine system?

2. Why are feedback controls important?

3. Name four endocrine glands, and tell what each one does in the body.

4. **Applying Concepts** Epinephrine, the fight-or-flight hormone, increases the level of glucose in the blood. Why would this be important in times of stress?

5. **Illustrating Concepts** Look around your house for an example of a feedback control. Draw a diagram explaining how this feedback control works to start and stop an action.

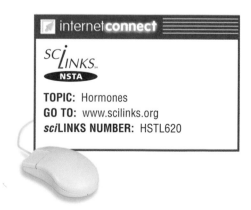

internet**connect**

SC*i*LINKS.
NSTA

TOPIC: Hormones
GO TO: www.scilinks.org
*sci*LINKS NUMBER: HSTL620

Skill Builder Lab

You've Gotta Lotta Nerve

Your skin has thousands of nerve receptors that detect sensations such as heat, cold, and pressure. Your brain is designed to filter out or ignore most of the input it receives from these skin receptors. If this were not the case, simply wearing clothes would trigger so many responses that you couldn't function.

Some areas of the skin, such as the back of your hand, are more sensitive than others. In this activity, you will map the receptors for heat, cold, and pressure on the back of your hand.

MATERIALS

- fine-point washable pens or markers
- metric ruler
- graph paper
- eyedropper
- very cold water
- facial tissue
- hot tap water
- dissecting pin with a small piece of cork or a small rubber stopper covering the sharp end

Procedure

1 Form a group of three. One group member will volunteer the back of his or her hand to be tested, one will do the testing, and the third will record the results. Check with your teacher to see if you may switch roles so that each group member may play each part.

2 Use a fine-point washable marker or pen and a metric ruler to mark off a 3 cm × 3 cm square on the back of the volunteer's hand. Draw a grid within the area, spacing the lines approximately 0.5 cm apart. You will have 36 squares in the grid when you are finished. Examine the photograph below to make sure you have drawn the grid correctly.

3 Mark off three 3 cm × 3 cm areas on the graph paper. Make a grid in each area exactly as you did on the back of the volunteer's hand. Label one grid "Cold," another grid "Hot," and the third grid "Pressure."

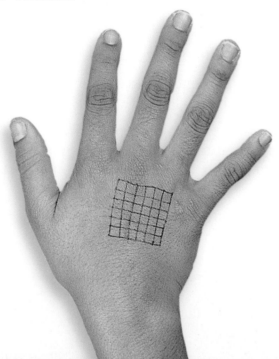

4 Begin locating receptors on the volunteer's hand. The volunteer should not look while his or her hand is being tested! Use the eyedropper to apply one small droplet of cold water on each square in the grid. The volunteer should tell you when he or she feels a cold droplet. On your graph paper, mark an *X* on the "Cold" grid in the square that corresponds to where the sensation of cold was felt on the hand. You will need to carefully blot the water off your partner's hand after several drops.

5 Repeat the test using hot-water droplets. The water will cool enough as it drops from the eyedropper that it will not hurt your partner. Mark an *X* on the "Hot" grid to indicate where the sensation of heat was felt on the hand.

6 Repeat the test using the head—not the point!—of the pin. Touch the skin with the pinhead to detect pressure receptors. Use a very light touch. Mark an *X* on the "Pressure" grid to indicate where pressure was felt on the hand.

Analysis

7 Count the number of *X*s in each grid. How many heat receptors are there per 3 cm^2? cold receptors? pressure receptors?

8 Do you have areas on the back of your hand where the receptors overlap? Why or why not?

9 How do you think the results of this experiment would be similar or different if you mapped an area of your forearm? the back of your neck? the palm of your hand?

10 Prepare a written report that includes a description of your investigation and a discussion of items 7–9.

Going Further
In the library or on the Internet, research what happens if a receptor is continuously stimulated. Does the kind of receptor make a difference? Does it make a difference how intense the stimulation is? Explain.

Chapter Highlights

Vocabulary

central nervous system (*p. 76*)

peripheral nervous system (*p. 76*)

neuron (*p. 77*)

impulse (*p. 77*)

receptor (*p. 78*)

nerve (*p. 78*)

brain (*p. 79*)

reflex (*p. 82*)

Section Notes

• The central nervous system includes the brain and spinal cord. The peripheral nervous system includes nerves and sensory receptors.

• A neuron receives information at branched endings called dendrites and passes information to other cells along a fiber called an axon.

• Sensory neurons detect information about the body and its environment. Motor neurons carry messages from the brain and spinal cord to other parts of the body.

• The cerebrum is the largest part of the brain and is involved with thinking, sensations, and voluntary muscle control.

• The cerebellum is the second largest part of the brain. It keeps track of the body's position and helps maintain balance.

• The medulla controls involuntary activities such as heart rate, blood pressure, and breathing.

• Pain signals can trigger a quick, involuntary action, called a reflex, in which a motor neuron sends a message to a muscle without first receiving a signal from the brain.

☑ Skills Check

Math Concepts

THE SPEED OF AN IMPULSE An impulse travels very fast. As shown in the MathBreak on page 77, to calculate the amount of time that it takes for an impulse to travel a certain distance, you must first know the speed it is traveling. Then you can divide the distance by the speed to get the time. For example, if an impulse travels 150 m/s, it would take it 0.02 seconds to travel 3 m.

$$\text{time} = \frac{3 \text{ m (distance)}}{150 \text{ m/s (speed)}} = 0.02 \text{ s}$$

Visual Understanding

PATH OF LIGHT Look back at Figure 10 on page 84 to review the path of light entering the eye. The light first passes through the transparent cornea, then through the opening called the pupil, and then through the lens. At the back of the eye, the light is detected by receptors in the retina.

SECTION 2

Vocabulary

retina *(p. 84)*

rods *(p. 84)*

cones *(p. 84)*

iris *(p. 85)*

lens *(p. 85)*

cochlea *(p. 86)*

Section Notes

- Different kinds of receptors in the skin are responsible for detecting touch, pressure, temperature, and pain.

- The retina of the eye contains photoreceptors that react to light and cause impulses to be sent to the brain.

- The lens of the eye can change shape to adjust the point of focus so that the image is focused on the retina. Improper focus can usually be corrected with glasses or contact lenses.

- Special receptors inside the cochlea of the ear react to sound waves and send impulses to the brain.

- Receptors for taste are located in taste buds on the bumps of the tongue.

- Receptors for smell are on olfactory cells located in the upper part of the nasal cavity.

SECTION 3

Vocabulary

endocrine system *(p. 88)*

gland *(p. 88)*

hormone *(p. 88)*

feedback control *(p. 90)*

Section Notes

- The endocrine system communicates with other systems using chemicals called hormones.

- Hormones are made in endocrine glands.

- The adrenal glands secrete hormones that help the body cope with stress. Epinephrine is the hormone most associated with fight-or-flight situations.

- Feedback control is the body's way of turning glands on and off so that they release hormones only when necessary.

 internet**connect**

GO TO: go.hrw.com

Visit the **HRW** Web site for a variety of learning tools related to this chapter. Just type in the keyword:

KEYWORD: HSTBD4

 *sci*LINKS℠ **NSTA**

GO TO: www.scilinks.org

Visit the **National Science Teachers Association** on-line Web site for Internet resources related to this chapter. Just type in the *sci*LINKS number for more information about the topic:

TOPIC: The Nervous System	*sci*LINKS NUMBER: HSTL605
TOPIC: The Senses	*sci*LINKS NUMBER: HSTL610
TOPIC: The Eye	*sci*LINKS NUMBER: HSTL615
TOPIC: Hormones	*sci*LINKS NUMBER: HSTL620

Chapter Review

To complete the following sentences, choose the correct term from each pair of terms listed below:

1. Your brain and spinal cord make up your __?__. (*central nervous system* or *peripheral nervous system*)

2. Sensory receptors in the __?__ detect vibrations. (*cochlea* or *eardrum*)

3. Epinephrine is produced by the adrenal glands in response to __?__. (*glucose* or *stress*)

4. The part of a neuron that passes an impulse to other cells is the __?__. (*dendrite* or *axon terminal*)

5. The medulla is mostly responsible for activities that are __?__. (*involuntary* or *voluntary*)

6. Receptors that can convert light into impulses are found in the __?__. (*olfactory cells* or *retina*)

UNDERSTANDING CONCEPTS

Multiple Choice

7. Which of the following has receptors for smelling?
   a. cochlea cells
   b. thermoreceptors
   c. olfactory cells
   d. optic nerve

8. Which of the following gives eyes their color?
   a. iris
   b. cornea
   c. lens
   d. retina

9. Which of the glands is associated with goiters?
   a. adrenal
   b. pituitary
   c. thyroid
   d. pancreas

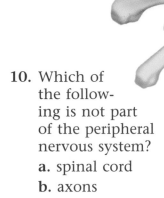

10. Which of the following is not part of the peripheral nervous system?
   a. spinal cord
   b. axons
   c. sensory receptors
   d. motor neurons

11. Which part of the brain regulates blood pressure?
   a. right cerebral hemisphere
   b. left cerebral hemisphere
   c. cerebellum
   d. medulla

12. Which of the following is associated with the endocrine system?
   a. reflex
   b. salivary gland
   c. fight-or-flight response
   d. voluntary response

Short Answer

13. Describe several situations in which your adrenal glands might release epinephrine, causing you to have a fight-or-flight reaction.

14. What causes the size of your pupils to change?

15. What is a reflex? How does a reflex enable you to act quickly?

16. What is the function of the middle-ear bones?

17. Using the terms you learned in this chapter, write down a step-by-step sequence for the path taken by an impulse, beginning at a pain receptor in your left big toe. Be sure to mention each kind of neuron and its parts as well as specific organs in the nervous system.

Concept Map

18. Use the following terms to create a concept map: the nervous system, spinal cord, medulla, peripheral nervous system, brain, cerebrum, central nervous system, cerebellum.

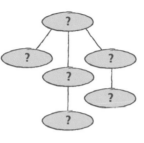

Look at the drawing below, and answer the following questions:

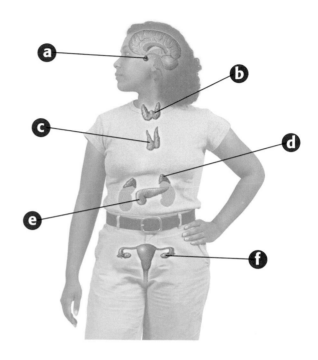

CRITICAL THINKING AND PROBLEM SOLVING

Write one or two sentences to answer the following questions:

19. Why is it important to have a lens that can change shape inside the eye?

20. Why can the nervous system have a faster effect on the body than the endocrine system?

21. Why is it important that reflexes occur without thought?

MATH IN SCIENCE

22. Sound travels about 335 m/s (1 km is equal to 1,000 m). How many kilometers would a sound travel in 1 minute?

23. Some axons can send one impulse every 0.4 milliseconds. One second is equal to 1,000 milliseconds. How many impulses could one of these axons send every second?

24. Which letter identifies the gland that regulates blood sugar?

25. Which letter identifies the gland that releases a hormone that stimulates the birth process in pregnant women?

26. Which letter identifies the gland that helps the body fight disease?

Reading Check-up

Take a minute to review your answers to the Pre-Reading Questions found at the bottom of page 74. Have your answers changed? If necessary, revise your answers based on what you have learned since you began this chapter.

Science, Technology, and Society

Light on Lenses

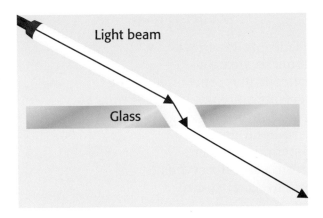

▲ *Light changes speed and direction when it passes from one material into another.*

Can you see in pitch darkness? No, of course not! You need light to see. But there is something else you need in order to see. You need a lens. A **lens** is a curved transparent object that *refracts,* or bends, light.

Lenses are necessary to focus light in all kinds of applications, including in telescopes, microscopes, binoculars, cameras, contact lenses, eyeglasses, and magnifying lenses.

Light Bounces

To learn how lenses work, you must first know something about how light travels. A ray of light travels in a straight path from its source until it strikes an object. When light strikes an object, much of the light bounces off, or is reflected. The light reflects from the object at the same angle that it struck the object in the first place.

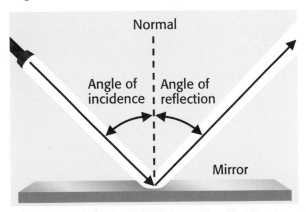

▲ *The angle formed by the incoming light (angle of incidence) always equals the angle of the reflected light (angle of reflection).*

Lenses Bend Light

A lens allows light to travel through it. However, as the light passes through the lens, it is refracted. **Refraction** is the bending of a light ray as it passes from one transparent material into another, such as when light traveling through air passes through a glass lens.

The type of lens determines how much and in which direction the light is bent. A lens that is thicker in the middle than at its edges is called a **convex lens.** This type of lens bends light toward its center. Convex lenses are used in magnifying glasses, microscopes, and telescopes. The lenses in your eyes are convex lenses.

A lens that is thinner in the middle than at its edges is called a **concave lens.** This type of lens bends light away from its center. Both convex lenses and concave lenses are often used to help correct vision. Convex lenses are also used in combination with concave lenses in cameras to focus light on the film.

Light Your Way

▶ Do some additional research to find out what a photorefractive keratectomy (PRK) is and how it works to correct a person's vision.

Eureka!

Pathway to a Cure

Do you know what would happen if your brain sent out too many impulses to the muscles in your body? First the overload would increase the number of contractions in your muscles, and it would be difficult to carry out simple movements, like scratching your arm or picking up a glass. Even when you wanted to rest, your muscles would continue to tremble. This is what happens to people with Parkinson's disease, and unfortunately, there is still no cure.

The Disease

Parkinson's disease affects the cells in the brain that regulate muscles. These cells require the chemical dopamine, which slows down the activity of nerves so they can function properly. But if the cells that supply the muscle-regulating cells with dopamine are damaged, the brain will send continuous impulses to the muscles. This results in Parkinson's disease.

Parkinson's disease is often diagnosed only after a person has already lost about 80 percent of his or her dopamine-supplying cells. Although there is no known cure for Parkinson's disease, some patients can be treated with chemicals that act like dopamine. Unfortunately, these substitutes are not as good as the real thing. Dopamine itself cannot be given because it cannot pass from the blood into the brain tissue.

Breakthrough

Dr. Bertha Madras studies the effects of drug addiction on the brain. While studying the effects of cocaine addiction, she discovered that a chemical called tropane attaches itself to the same nerves that release dopamine in the brain. This discovery may be used to detect and diagnose Parkinson's disease earlier and at a lower cost to the patient.

A Glow in the Darkness

Madras and her colleagues thought they could use tropane to study the cells that release dopamine. They added a radioactive component to the tropane to make a chemical called altropane. Altropane also attaches to the dopamine-releasing cells. But unlike tropane, altropane glows, so it shows up in a brain scan. Healthy people have large areas where the altropane attaches. Among patients with Parkinson's disease, because of the nerve loss, the altropane attaches to fewer nerves. Therefore, brain scans from these patients do not have as many glowing collections of altropane.

Using this new procedure to diagnose Parkinson's disease could allow doctors to find the disease in people before the neurons are severely damaged or completely lost.

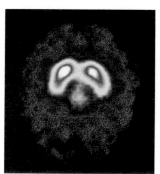 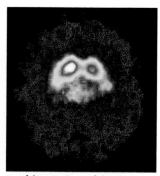

Healthy subject   Parkinson's subject

▲ *Brain scans, such as the ones above, can be used to diagnose Parkinson's disease.*

Activity

▶ Find out what a Single Photon Emission Computed Tomography (SPECT) image is and how it is used to study Parkinson's disease.

CHAPTER

5

Reproduction and Development

Pre-Reading
Questions

1. Do all animals have two
parents?

2. What makes you physi-
cally different from an
adult?

3. What percentage of genes
do you inherit from your
mother? your father?

SNEAK PREVIEW

If someone had taken your picture when your mother was
about seven months pregnant with you, it would have
looked very much like this photograph. By the eighth
month, your eyes opened and you could see light. Can
you believe how much you changed in such a short time?
In this chapter, you will learn about how a single cell
grows and develops into a complete person. You will
also learn how you continue to change from infancy
through adulthood.

HOW GROWS IT?

As you read this, you are aging. Your body is growing into the body of an adult. But does your body have the same proportions that an adult's body has? Do this exercise to find out.

Procedure

1. Have a classmate help you measure your total height, head height, and leg length with a **tape measure** and **meterstick.** Your teacher will tell you how to take these measurements.

2. Calculate your head-to-body proportion and leg-to-body proportion. Use the following equations:

$$\text{head proportion} = \left(\frac{\text{head height}}{\text{body height}}\right) \times 100$$

$$\text{leg proportion} = \left(\frac{\text{leg length}}{\text{body height}}\right) \times 100$$

3. Your teacher will give you the head, body, and leg measurements of three adults. Calculate their proportions. Record all the measurements and calculations.

Analysis

4. Using the direct evidence you collected, evaluate how your proportions compare with the proportions of adults.

Terms to Learn

asexual reproduction

sexual reproduction

egg

sperm

zygote

external fertilization

internal fertilization

What You'll Do

◆ Distinguish between asexual and sexual reproduction.

◆ Explain the difference between external and internal fertilization.

◆ Describe the three different types of mammalian development.

Animal Reproduction

The life span of some living things is very short compared with ours. For instance, a fruit fly lives only about 80 days. Other organisms live for a long time. A bristlecone pine can live for 2,000 to 6,000 years. But all living things eventually die. If a species is to survive, its members must reproduce.

A Chip off the Old Block

Some animals, particularly simpler ones, reproduce asexually. In **asexual reproduction,** a single parent has offspring that are genetically identical to itself.

One kind of asexual reproduction is called *budding.* This occurs when a small part of the parent's body develops into an independent organism. The hydra shown in **Figure 1** is reproducing asexually by budding. The young hydra is genetically identical to its parent.

Fragmentation is another type of asexual reproduction. In fragmentation, an organism breaks into two or more parts, each of which may grow into a separate individual. Sea stars can reproduce by fragmentation. Because sea stars eat oysters, people used to try to kill sea stars by chopping them into pieces and throwing the pieces back into the water. They didn't know that each arm of a sea star can grow into an entire organism! This can be seen in **Figure 2.**

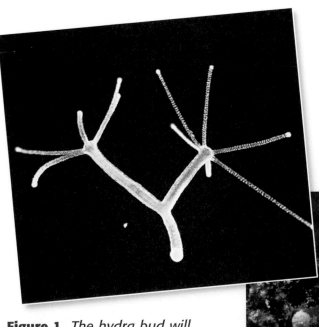

Figure 1 *The hydra bud will separate from its parent. Buds from other organisms, such as coral, remain attached to the parent.*

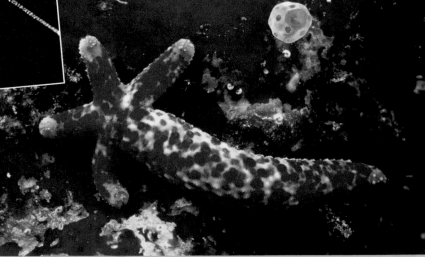

Figure 2 *The largest arm on this sea star was a fragment, from which the rest of the sea star has grown. In time, all of the sea star's arms will grow to the same size.*

It Takes Two

Sexual Reproduction produces offspring by combining the genetic material of more than one parent. Most animals, including humans, reproduce sexually. Sexual reproduction most commonly involves two parents, a male and a female. The female parent produces sex cells called **eggs.** The male parent produces sex cells called **sperm.** When an egg's nucleus joins with a sperm's nucleus, a new kind of cell, called a **zygote,** is created. This joining of an egg and sperm is known as *fertilization.*

Review of Meiosis Genes are located in *chromosomes.* All human cells except egg and sperm cells contain 46 chromosomes. Eggs and sperm each contain only 23 chromosomes. Eggs and sperm are formed by a process known as *meiosis.*

In humans, meiosis involves the division of one cell with 46 chromosomes into four sex cells with 23 chromosomes each. When an egg and sperm join to form a zygote, the original number of 46 chromosomes is restored. This combination of genes from the father and mother results in a zygote that will grow into a unique individual. **Figure 3** shows how genes are intermixed through three generations.

÷ 5 ÷ Ω ≤ ∞ +Ω √ 9 ∞≤ Σ 2
+

MATH **BREAK**

Chromo-Combos

A cell undergoes meiosis. This cell has 6 chromosomes in 3 pairs. How many chromosomal combinations are possible in the formed sex cells? To find out, use the following formula:

2^x = possible variations where x = the number of pairs.

2^3 (or $2 \times 2 \times 2$) = 8 Therefore, 8 variations are possible.

A typical human cell has 46 chromosomes in 23 pairs. If the cell undergoes meiosis, how many chromosomal combinations are possible in the resulting sex cells?

Figure 3 *Eggs and sperm contain genes. You inherit genes from both of your parents. Your parents each inherited genes from both of their parents.*

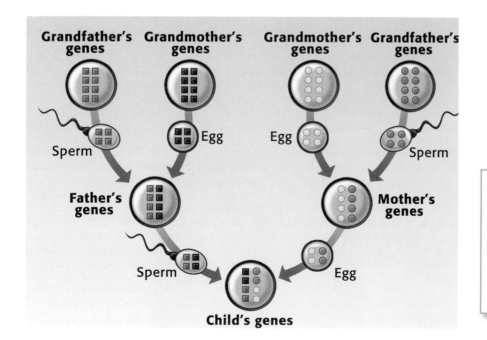

Grandfather's genes Grandmother's genes Grandmother's genes Grandfather's genes

Sperm Egg Egg Sperm

Father's genes Mother's genes

Sperm Egg

Child's genes

✓ Self-Check

What is the difference between sexual and asexual reproduction? *(See page 212 to check your answer.)*

Internal and External Fertilization

Depending on the animal, fertilization may occur either outside or inside the female's body. Some fishes and amphibians reproduce by **external fertilization,** in which the sperm fertilize the eggs outside the female's body. External fertilization must take place in a moist environment so the delicate zygotes won't dry out.

Many frogs, such as those pictured in **Figure 4,** mate every spring. The female frog releases her eggs first. The male frog then releases sperm over the eggs to fertilize them. The frogs leave the fertilized eggs to develop on their own. In about two weeks, the eggs hatch into tadpoles.

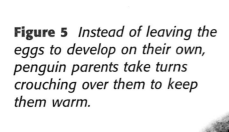

Figure 4 *Frogs fertilize their eggs externally. Some species can produce more than 300 offspring in one season.*

The Inside Story With **internal fertilization,** eggs and sperm join inside the female's body. Reptiles, birds, mammals, and some fishes reproduce by internal fertilization. Many animals that use internal fertilization lay fertilized eggs. The female penguin in **Figure 5,** for example, usually lays one or two eggs after internal fertilization has occurred.

In most mammals, internal fertilization is followed by the development of a fertilized egg inside the mother's body. Many mammals give birth to young that are well developed. Young zebras, like the one in **Figure 6,** can stand up and nurse almost immediately after birth.

Figure 5 *Instead of leaving the eggs to develop on their own, penguin parents take turns crouching over them to keep them warm.*

Figure 6 *This zebra has just been born, but he is already able to stand. Within an hour, he will be able to run.*

Making Mammals

All mammals reproduce sexually and nurture their young with milk. There are some differences in how mammals produce off- spring, but every mammal follows one of three types of development.

Monotremes   Mammals that lay eggs are *monotremes*. Two families of monotremes live today—the echidna and the platypus. After these animals lay their eggs, there is an incubation period that lasts up to 2 weeks. When the eggs hatch, the babies are very unde- veloped. They crawl into a fold of their mother's skin and are nourished by the milk that oozes from her pores.

Marsupials   Mammals that give birth to live young that are only partially developed are *marsupials*. There are about 260 species of marsupials. Most of them have pouches where their young develop, but some South American species do not have this feature. Marsupials with pouches have extra bones to help support the weight of their young, as can be seen in **Figure 7.** When a baby marsupial attaches itself to its mother's nipple, the nipple expands in the baby's mouth to prevent the baby from separating from its mother.

Figure 7 *The skeleton of this opposum has two extra bones extending forward from its pelvis to help support the weight of its young.*

Placental Mammals   There are almost 4,000 different species of placental mammals. These include whales, elephants, armadillos, bats, horses, and humans. *Placental mammals* nour- ish their young internally before birth. Newborn placental mammals are highly developed compared with newborn marsupials or monotremes.

SECTION REVIEW

1. How many parents are needed to reproduce asexually?

2. What is the difference between monotremes and marsupials?

3. How is a zygote formed?

4. **Applying Concepts**   Birds lay eggs, but they are not con- sidered monotremes. Explain why.

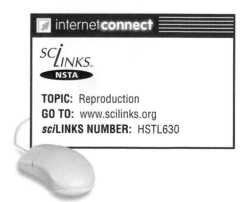

internet**connect**

SC*i*LINKS
NSTA

TOPIC: Reproduction
GO TO: www.scilinks.org
*sci*LINKS NUMBER: HSTL630

Terms to Learn

testes	ovaries
puberty	fallopian tube
vas deferens	uterus
semen	vagina
penis	infertile

What You'll Do

◆ Describe the functions of the male and female reproductive systems.

◆ Discuss disorders and diseases that are associated with human reproduction.

Human Reproduction

When a human sperm and egg combine, a new human begins to grow. About 9 months later, a mother gives birth to her baby. But what happens before that? Where do eggs and sperm come from?

The Male Reproductive System

The male reproductive system, shown in **Figure 8,** produces sperm and delivers it to the female reproductive system. The **testes** (singular, *testis*) make sperm and testosterone. Testosterone is the principal male sex hormone. It regulates the production of sex cells and the development of male characteristics.

Sperm Production The human body is usually around 37°C, but sperm cannot develop properly at such high temperatures. That is why the two testes rest in the *scrotum,* a skin-covered sac that hangs from the body. The scrotum is about 2 degrees cooler than the body. Inside each testis are masses of tightly coiled tubes called *seminiferous* (SEM uh NIF uhr uhs) *tubules* (TOO BYOOLZ), where sperm are produced. A healthy adult male produces several hundred million sperm each day! This massive, continuous sperm production begins at puberty. **Puberty** is the time of life when the sex organs of both males and females become mature.

Before sperm leave a testis, they are stored in a tube called an *epididymis* (EP uh DID i mis). Another tube called a **vas deferens** (vas DEF uh RENZ) passes from each epididymis into the body. As sperm swim through the vas deferens, they mix with fluids from several glands. The mixture of sperm and fluids is called **semen.**

To leave the body, semen passes through the vas deferens into the *urethra,* the tube that runs through the penis. The **penis** transfers semen into the female's body during sexual intercourse.

Figure 8 The Male Reproductive System

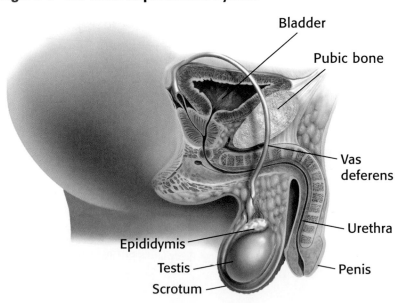

Bladder

Pubic bone

Vas deferens

Urethra

Epididymis

Testis

Scrotum

Penis

The Female Reproductive System

The female reproductive system, shown in **Figure 9,** produces eggs, nurtures fertilized eggs, and gives birth. The **ovaries** produce the eggs. The two ovaries also produce sex hormones, such as estrogen and progesterone, that regulate the release of eggs and direct the development of female characteristics.

The Egg's Journey An ovary contains eggs in various stages of development. As an egg matures, it becomes a huge cell, growing to almost 200,000 times the size of a sperm. During *ovulation,* an egg is ejected through the ovary wall. Then the egg passes into a fallopian (fuh LOH pee uhn) tube. A **fallopian tube** leads from each ovary to the uterus. The **uterus** is the organ where a baby grows and develops.

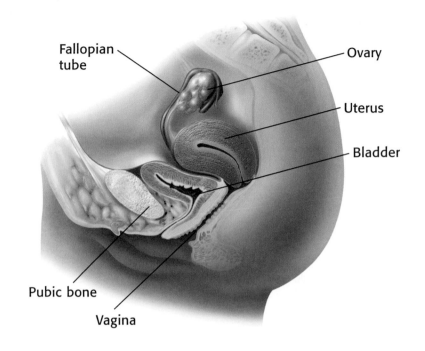

Figure 9 The Female Reproductive System

Fallopian tube · Ovary · Uterus · Bladder · Pubic bone · Vagina

Every month starting at puberty, the lining of the uterus thickens in preparation for pregnancy. If fertilization occurs, the zygote moves down a fallopian tube and embeds in the lining of the uterus. When a baby is born, it passes from the uterus through the vagina. The **vagina** is the same passageway that received the sperm during sexual intercourse.

Menstrual Cycle To prepare for pregnancy, a female's reproductive system goes through several changes. These changes, called the menstrual cycle, usually occur every 28 days. The first day of the cycle is the beginning of *menstruation,* the monthly discharge of blood and tissue from the uterus. Menstruation lasts about 5 days. As soon as menstruation is over, the uterus's lining begins to build up again in preparation for ovulation. Ovulation typically occurs around the 14th day of the cycle. If the egg isn't fertilized by the time it reaches the uterus, it will deteriorate. Menstruation will flush the egg away, starting the cycle over again. A female's menstrual cycle begins at puberty and continues until late middle age.

MATH BREAK

Counting Eggs

1. The average human female ovulates every month from about age 12 to about age 50. How many mature eggs can she produce during that time period?

2. A female's ovaries typically house 2 million immature egg cells. If she ovulates regularly from age 12 to age 50, what percentage of her eggs will mature?

Irregularities and Disorders

In most cases, the human reproductive system completes its functions flawlessly. However, as with any body system, there can sometimes be irregularities or disorders.

Figure 10 *Identical twins have the exact same genes. Many identical twins who are reared apart have similar personalities and interests.*

Multiple Births Have you ever seen a pair of identical twins? Sometimes they are so similar that even their parents can't tell them apart. About one pair of identical twins is born for every 250 births. Another type of twins, called fraternal twins, is also born frequently. Fraternal twins can look very different from each other.

Twins, such as those shown in **Figure 10,** are the most common type of multiple births, but humans can also have triplets (3 babies), quadruplets (4 babies), quintuplets (5 babies), and so on. These types of multiple births are extremely rare. For instance, quadruplets occur only about once in every 705,000 births. Do you know what circumstances result in a multiple birth? To find out, do the Apply exercise at the bottom of this page.

Ectopic Pregnancy In a normal pregnancy, the fertilized egg travels to the uterus and attaches itself to the uterus's wall. In an *ectopic* (ek TAHP ik) *pregnancy,* the fertilized egg attaches itself to a fallopian tube or another area of the reproductive system. Because the zygote cannot develop correctly outside of the uterus, an ectopic pregnancy can be very dangerous for both the mother and child.

Two Types of Twins

Zach and Drew are fraternal twins. They don't look much alike. Emily and Carol are identical twins. They are hard to tell apart. Why are some twins identical and others fraternal? Consider the two possibilities illustrated at right: In *A,* the mass of cells from a single fertilized egg separates into two halves early in development, and in *B* two eggs are released by an ovary and fertilized by two different sperm cells. Record the answers to the following questions in your ScienceLog:

1. Which instance, *A* or *B,* would produce identical twins? Explain your answer.
2. Could fraternal twins be (a) both boys, (b) both girls, (c) one girl and one boy, or (d) all of the above?

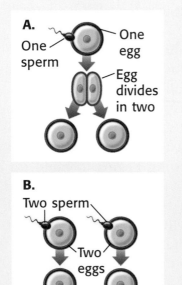

Infertility In the United States, about 15 percent of married couples have difficulty producing offspring. Many of these couples are **infertile,** which means they are unable to have children. Men may be infertile because they cannot produce enough healthy sperm. This is called a low sperm count. Women may be infertile because they do not ovulate normally. Sexually transmitted diseases can also cause infertility.

STDs *Sexually transmitted diseases* (STDs) are diseases that can pass from an infected person to an uninfected person during sexual contact. An STD you may have heard about is the acquired immune deficiency syndrome (AIDS). Other common STDs are shown in the table below.

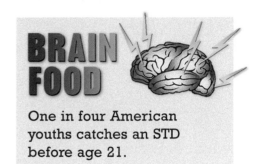

BRAIN FOOD

One in four American youths catches an STD before age 21.

The Spread of STDs in the U.S.	
STD	**Approximate new cases each year**
Chlamydia	3–10 million
Gonorrhea	1–3 million
Genital warts	1 million
AIDS	750,000
Genital herpes	500,000
Syphilis	120,000

Chemistry CONNECTION

Many chemicals in pollutants are similar to female hormones. Studies are beginning to link these chemicals with early menstruation and low sperm counts.

Cancer Cancer, the uncontrolled division of cells, sometimes occurs in the reproductive organs. The testes and the prostate gland, a gland that produces the fluid in semen, are common sites of cancer in men over age 50. In women, the ovaries and breasts are common sites of cancer.

SECTION REVIEW

1. What is the difference between sperm and semen?

2. Can a woman become pregnant at any time of the month? Explain.

3. Define *sexually transmitted diseases,* and give three examples.

4. **Applying Concepts** How are the ovaries similar to the testes? How are they different?

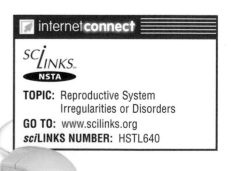

internetconnect

SCiLINKS
NSTA

TOPIC: Reproductive System Irregularities or Disorders
GO TO: www.scilinks.org
*sci*LINKS NUMBER: HSTL640

Growth and Development

Every one of us starts out as a single cell that will become a complete person. We are made of millions of cells, each with its own job to do. You, of course, are no exception. You have become a very complex individual, capable of thousands of different thoughts and actions. It is hard to believe that a person as remarkable as you began your life as a single cell, but that is just what happened.

A New Life

The natural process of creating a human baby starts when a man deposits millions of sperm into a woman's vagina during sexual intercourse. Most of the sperm will die because of the vagina's acidic environment, but a few hundred are able to make it through the uterus and into the fallopian tube, as can be seen in **Figure 11.** The surviving sperm cover the egg, releasing enzymes that help dissolve the egg's outer covering. As soon as one sperm gets through, a membrane closes around the fertilized egg. This membrane keeps other sperm cells from entering.

Figure 11   Fertilization and Implantation

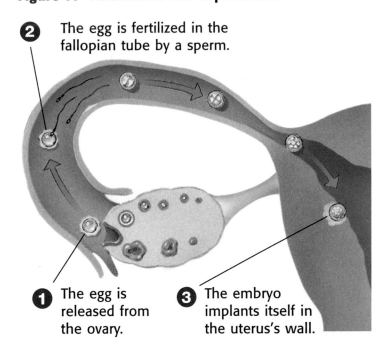

② The egg is fertilized in the fallopian tube by a sperm.

① The egg is released from the ovary.

③ The embryo implants itself in the uterus's wall.

Implantation   The fertilized egg travels down the woman's fallopian tube toward her uterus. The journey takes about 5 days. The zygote undergoes cell division many times during the trip. By the time it reaches the uterus, it is a tiny ball of cells called an **embryo.** During the next few days, the embryo must embed itself in the thick, nutrient-rich lining of its mother's uterus. This process is called **implantation,** and only about 30 percent of all embryos successfully do it. **Figure 12** shows an implanted embryo.

The embryo's actual size is slightly smaller than the period at the end of this sentence.

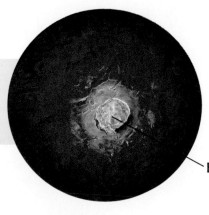

Figure 12 *This embryo has implanted in the wall of its mother's uterus.*

Embryo

Before Birth

When the embryo implants itself in a woman's uterus, the woman is officially pregnant. For the embryo to survive, a special two-way exchange organ called a **placenta** begins to grow. The placenta contains a network of blood vessels that provide the embryo with oxygen and nutrients from the mother's blood. Wastes that the embryo produces are removed by the placenta and transported to the mother's blood for her to excrete. Although the embryo's blood and the mother's blood flow very near each other inside the placenta, they never actually mix.

Self-Check

Why is it important that the embryo be implanted in the uterus and not elsewhere? *(See page 212 to check your answer.)*

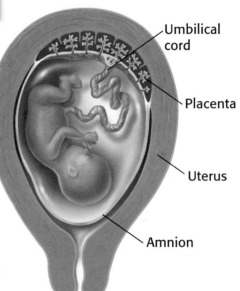

Figure 13 *The placenta, amnion, and umbilical cord are the life support system for the fetus.*

First Month About 1 week after implantation, the embryo's blood cells and a heart tube form. Then the heart tube begins to twitch, starting the rhythmic beating that will continue for the individual's entire life. By the fourth week, the embryo is almost 2 mm long. Surrounding the embryo is a thin, fluid-filled membrane called the *amnion*, which is formed to protect the growing embryo from shocks. The **umbilical cord** is another new development. It connects the embryo to the placenta. The umbilical cord, amnion, and placenta can be seen in **Figure 13.**

Second Month By the time the embryo is 4 weeks old, it has the beginnings of a brain and spinal cord. It also has tiny limb buds that will eventually develop into arms and legs. Its nostrils, eyelids, hands, and feet then begin to form. Its muscles begin to develop, and for the first time in its life, its brain begins to send signals to other parts of its body. Despite all these transformations, the embryo is still only about the size of a peanut. **Figure 14** shows a 5-week-old embryo.

Figure 14 A 5-Week-Old Embryo

Actual size

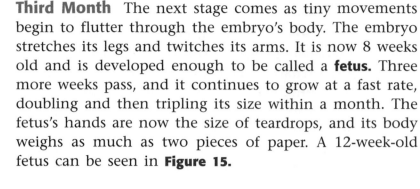

Figure 15 A 12-Week-Old Fetus

Actual size

Third Month The next stage comes as tiny movements begin to flutter through the embryo's body. The embryo stretches its legs and twitches its arms. It is now 8 weeks old and is developed enough to be called a **fetus.** Three more weeks pass, and it continues to grow at a fast rate, doubling and then tripling its size within a month. The fetus's hands are now the size of teardrops, and its body weighs as much as two pieces of paper. A 12-week-old fetus can be seen in **Figure 15.**

Fourth to Sixth Month The fetus's 13th week of life passes, and suddenly new movement! It can blink its eyes for the first time, swallow, hiccup, make a fist, and curl its toes that now have tiny nails. By the fourth month, the fetus starts to make even bigger movements. The mother now knows when her baby kicks its legs or stretches its arms.

During the fifth month, the fetus is about 20 cm long. Taste buds form on its tongue, and eyebrows form on its face. The fetus begins to hear sounds through the wall of its mother's uterus. Look at the timeline in **Figure 16** to review the changes that take place in the fetus.

Figure 16 Pregnancy Timeline

Weeks of the First Three Months	
1 and 2	The egg is fertilized by a sperm. The fertilized egg makes its way to the uterus, where it burrows into the lining. The fertilized egg is now called an embryo.
3 and 4	Most major organ systems have started to form. The heart starts to beat around day 22. The placenta is completely formed by the fourth week.
5 and 6	Facial features begin to take shape. The skeleton begins to form.
7 and 8	Muscle movement begins. The embryo is now called a fetus.
9 and 10	Arms, legs, hands, and feet have formed.
11 and 12	The internal organs have formed.

Weeks of the Second Three Months	
13 and 14	The circulatory system is working.
15 and 16	The mother may start to feel the fetus move.
17 and 18	The fetus responds to sound.
19 and 20	The fetus is now about 20 cm long.
21 and 22	
23 and 24	Eyelashes and eyebrows appear.

Weeks of the Third Three Months	
25 and 26	The eyes open.
27 and 28	The fetus can "practice breathe."
29 and 30	Layers of fat form beneath skin.
31 and 32	
33 and 34	Organs are fully functional.
35 and 36	The fetus responds to light.
Birth	The baby is born.

Seventh to Ninth Month The seventh month is when the fetus's memories begin to form. During this time, its lungs start to "practice breathe," moving up and down continuously as if breathing real air. If the fetus's mother smokes one cigarette during this stage, the fetus's lung movement will stop for up to an hour. The fetus in **Figure 17** is starting its first lung movement.

By the eighth month, the fetus's open eyes can perceive light through its mother's abdominal wall, and its sleeping pattern starts to be influenced by sunlight. When the fetus is asleep, it dreams. Can you imagine what its dreams might be about?

Figure 17 A 21-Week-Old Fetus

Actual size of hand

Birth

After about 9 months, the fetus is ready to live outside of its mother. The mother goes through a series of muscular contractions called *labor*. During labor, the fetus is usually squeezed headfirst through the vagina. There is little room to spare, and the fetus's head is temporarily squashed out of shape as the fetus passes through its mother's pelvis. Suddenly bright lights and cold air surround the newborn baby. It gasps, fills its lungs with air for the first time, and cries.

The baby in **Figure 18** is still connected to the placenta by its umbilical cord. The doctor or midwife assisting the mother ties and cuts the umbilical cord. The baby's navel is all that will remain of the point where the umbilical cord was attached. After the mother expels the placenta from her body, labor is complete.

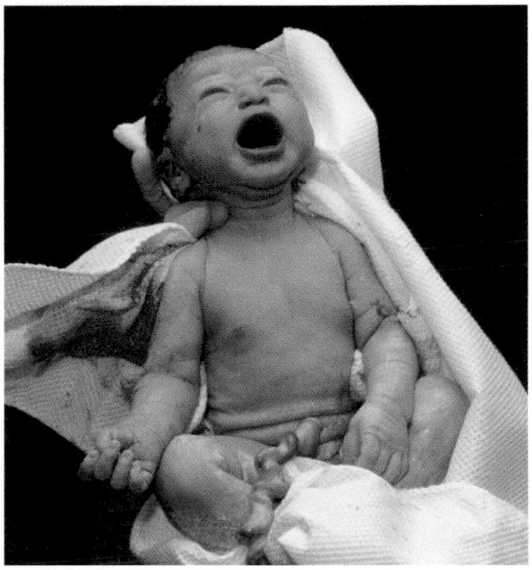

Figure 18 *This newborn baby is still attached to its umbilical cord. The average mass of a newborn baby is 3.3 kg. The average length is 50 cm.*

From Birth to Death

Of all the animals on this planet, humans have one of the longest life spans. Human infancy lasts 2 years—the same time it takes for most rabbits to be born, grow old, and die. Our childhood extends over a full decade, longer than many cats or dogs live. Humans can live for more than 100 years!

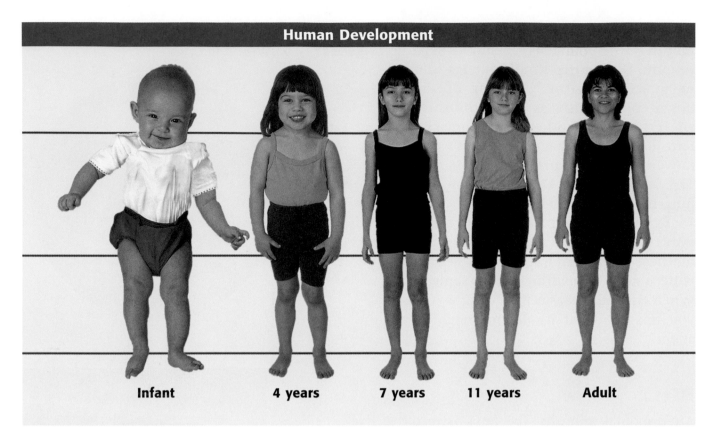

| Infant | 4 years | 7 years | 11 years | Adult |

Figure 19 *Five different stages of development are shown the same size so you can see how body proportions change as a person develops.*

Activity

Create a poster or timeline illustrating the different stages of human growth.

TRY at HOME

Infancy What life stages have you gone through since you were born? You have probably gone through most of the stages shown in **Figure 19.** You were an infant from birth to 2 years of age. During this time, you grew rapidly. Your teeth began to appear. You also became more coordinated as your nervous system developed. This enabled you to begin to walk.

Childhood Your childhood extends from 2 years to puberty. This is also a period of rapid growth. Your first set of teeth were slowly shed and replaced by permanent teeth. Your muscles became more coordinated, allowing you to do activities such as riding a bicycle and jumping rope. Your intellectual abilities also developed during this time.

Adolescence You are considered an adolescent from puberty to adulthood. During puberty, the reproductive systems of young males and females become mature. Puberty occurs in most boys sometime between the ages of 11 and 16. The young male body becomes more muscular, the voice becomes deeper, and body and facial hair appear. In most girls, puberty occurs between the ages of 9 and 14. During puberty in females, the amount of fat in the hips and thighs increases, the breasts enlarge, and body hair appears in areas such as the armpits. At this time, the young female also begins to menstruate.

Adulthood From about age 20 to age 40, you will be considered a young adult. You will be at the peak of your physical development. Beginning around age 30, certain changes associated with aging begin. The changes will be gradual and slightly different for everyone. Some of the early signs of aging include decreasing muscle flexibility, deteriorating eyesight, increasing body fat, and increasing hair loss.

The aging process will continue in a middle-aged adult (someone between 40 and 65). During this period, hair may become gray, athletic abilities will decline, and skin will wrinkle. Any person over 65 years old is considered an older adult. Although aging persists during this period of an individual's life, older adults can still lead active lives. Some of this country's most productive citizens are older adults, as can be seen in **Figure 20.**

Figure 20 *John Glenn, the first American to orbit Earth, returned to space at the age of 77.*

QuickLab

Life Grows On

Use Figure 19 on the previous page to complete this activity.

1. Use a **ruler** to measure the infant's head height. Then measure the infant's entire body height, including the head.

2. Calculate the percentage of the infant's head height to the infant's total height.

3. Repeat these measurements and calculations for the other stages shown in the figure.

Answer the following question in your ScienceLog:

As a baby grows into an adult, does the head grow faster or slower than the rest of the body? Why do you think this is so?

TRY at HOME

SECTION REVIEW

1. What is the difference between an embryo and a fetus?

2. Why does a membrane enclose an egg once a sperm has entered?

3. What developmental changes take place from birth to puberty?

4. **Applying Concepts** When astronauts work in space, they are sometimes attached to the spacecraft by a line called an umbilical. Why do you think the line has been given this name?

internet**connect**

sciLINKS
NSTA

TOPIC: Before Birth,
Growth and Development
GO TO: www.scilinks.org
*sci***LINKS NUMBER:** HSTL635, HSTL645

Skill Builder Lab

It's a Comfy, Safe World!

Before human babies are born, they lead a comfy life. By the seventh month, they just lie around sucking their thumb, blinking their eyes, and perhaps even dreaming. Most mammal babies develop within their mother's uterus, where they are surrounded by fluid and a placenta. Baby birds live inside a hard, protective shell until the baby has used up all the food supply. Is the internal environment in a placental mammal safer than a baby bird's environment? In this activity, you will create a model of a placental mammal's uterus to see how well it protects a fetus.

MATERIALS

- sealable plastic bags
- water
- mineral oil, cooking oil, syrup, or other thick liquid to represent fluid surrounding the fetus
- cotton, soft fabric, or other soft materials
- 3–5 soft-boiled eggs
- protective gloves

Procedure

1 Brainstorm several ideas about how you will construct and test your model. A peeled, soft-boiled egg will represent the fetus in your mammalian model. Review the structure of a uterus and amnion. Then build your model.

2 Using a computer or graph paper, make a data table similar to the "First Model Test" table below. Test your model, examine the egg for damage, and record your results.

3 If your soft-boiled egg broke during the first round of tests, you may need to modify your design. Remember to use your knowledge of a human uterus and amnion to help you improve your model. Repeat step 2, and record your results.

4 When you are satisfied with the design of your model, obtain another peeled, soft-boiled egg and an egg in the shell.

First Model Test	
Original model	**Modified model**
	DO NOT WRITE IN BOOK

Final Model Test	
	Test results
Model	
Egg in shell	

5 Make another data table similar to the "Final Model Test" table above. Repeat step 2 with your new eggs. Record your results in your data table. Use your table to organize, examine, and evaluate your data.

Analysis

6 Explain any differences in the test results for the mammalian model and the egg in the shell.

7 Use the direct evidence you gathered to evaluate which modification to your model was the most effective at protecting your fetus.

8 How well did your model represent a natural uterus? Identify the limitations of your model. Make recommendations for improving it.

Going Further
Compare the development of placental mammals with the development of marsupial mammals and monotremes.

Chapter Highlights

Vocabulary

asexual reproduction (*p. 102*)

sexual reproduction (*p. 103*)

egg (*p. 103*)

sperm (*p. 103*)

zygote (*p. 103*)

external fertilization (*p. 104*)

internal fertilization (*p. 104*)

Section Notes

- During asexual reproduction, a single parent produces offspring that are genetically identical to the parent. Budding and fragmentation are examples of asexual reproduction.

- During sexual reproduction, there is a union of an egg and a sperm.

- Each egg and sperm is the product of meiosis and contains half the usual number of chromosomes. The usual number of chromosomes is restored in the zygote.

- Sperm fertilize eggs outside the female's body in external fertilization. Sperm fertilize eggs inside the female's body in internal fertilization.

- Monotremes are egg-laying mammals. Marsupials are mammals that give birth to partially developed young. Placentals are mammals that give birth to well-developed young.

Vocabulary

testes (*p. 106*)

puberty (*p. 106*)

vas deferens (*p. 106*)

semen (*p. 106*)

penis (*p. 106*)

ovaries (*p. 107*)

fallopian tube (*p. 107*)

uterus (*p. 107*)

vagina (*p. 107*)

infertile (*p. 109*)

☑ Skills Check

Math Concepts

EGGS IN EXILE A woman does not ovulate while she is pregnant. Therefore, if a woman has three children, she will release at least 27 fewer eggs from her ovaries than she would if she never became pregnant.

3 children × 9 months of pregnancy = 27 eggs

Visual Concepts

MALE AND FEMALE REPRODUCTIVE SYSTEMS
The diagrams on pp. 106 and 107 show the male and female reproductive systems. Take another look at them, and make sure you recognize all the structures. Also note the similarities between the two systems. For instance, the ovaries have a similar function to the testes, and the fallopian tubes have a similar function to the vas deferens.

SECTION 2

Section Notes

- The male reproductive system produces sperm and delivers it to the female reproductive system. Sperm are produced in the seminiferous tubules and stored in the epididymis. Sperm leave the body through the urethra.

- The female reproductive system produces eggs, nourishes the developing embryo, and gives birth. An egg leaves one of two ovaries each month and travels to the uterus. If the egg is not fertilized, it disintegrates and menstruation occurs.

- Reproductive system disorders include infertility, cancer, and sexually transmitted diseases.

SECTION 3

Vocabulary

embryo *(p. 110)*

implantation *(p. 110)*

placenta *(p. 111)*

umbilical cord *(p. 111)*

fetus *(p. 112)*

Section Notes

- Fertilization occurs in a fallopian tube. From there, the zygote travels to the uterus and implants itself in the uterus's wall.

- After implantation, the placenta develops. The umbilical cord connects the embryo to the placenta. The amnion surrounds and protects the embryo.

- The embryo grows, developing limbs, nostrils, eyelids, and other features. By the eighth week, the embryo is developed enough to be called a fetus.

- Human life stages are infant (birth to 2 years), child (2 years to puberty), adolescent (puberty to 20 years), young adult (20 to 40 years), middle-aged adult (40 to 65 years), and older adult (older than 65 years).

Labs

My, How You've Grown!
(p. 182)

internet**connect**

GO TO: go.hrw.com

Visit the **HRW** Web site for a variety of learning tools related to this chapter. Just type in the keyword:

KEYWORD: HSTBD5

N S T A

GO TO: www.scilinks.org

Visit the **National Science Teachers Association** on-line Web site for Internet resources related to this chapter. Just type in the *sci*LINKS number for more information about the topic:

TOPIC: Reproduction	*sci***LINKS NUMBER:** HSTL630
TOPIC: Before Birth	*sci***LINKS NUMBER:** HSTL635
TOPIC: Reproductive System Irregularities or Disorders	*sci***LINKS NUMBER:** HSTL640
TOPIC: Growth and Development	*sci***LINKS NUMBER:** HSTL645

Chapter Review

To complete the following sentences, choose the correct term from each pair of terms listed below:

1. Reptiles, birds, and mammals reproduce sexually by __?__. (*internal fertilization* or *external fertilization*)

2. Sperm are produced in the __?__ within the testes. (*epididymis* or *seminiferous tubules*)

3. The sperm-containing fluid that exits the male body is known as __?__. (*semen* or *amniotic fluid*)

4. The release of an egg from the ovary occurs once each month and is called __?__. (*ovulation* or *menstruation*)

5. The organ of exchange between the developing embryo and the mother is the __?__. (*amnion* or *placenta*)

UNDERSTANDING CONCEPTS

Multiple Choice

6. The sea star can reproduce asexually by
   a. fragmentation.
   b. budding.
   c. external fertilization.
   d. internal fertilization.

7. The correct path of sperm through the male reproductive system is
   a. testes → epididymis → urethra → vas deferens.
   b. epididymis → urethra → testes → vas deferens.
   c. testes → vas deferens → epididymis → urethra.
   d. testes → epididymis → vas deferens → urethra.

8. If the first day of the menstrual cycle is the onset of menstruation, on what day does ovulation typically occur?
   a. 2nd day   c. 14th day
   b. 5th day   d. 28th day

9. Monotremes are different from placental mammals because they
   a. are mammals.
   b. have hair.
   c. nurture their young with milk.
   d. lay eggs.

10. All of the following are sexually transmitted diseases *except*
   a. chlamydia.   c. infertility.
   b. AIDS.   d. genital herpes.

11. Fertilization occurs in the __?__, and implantation occurs in the __?__.
   a. uterus, fallopian tube
   b. fallopian tube, vagina
   c. uterus, vagina
   d. fallopian tube, uterus

Short Answer

12. What human reproductive organs produce sperm? egg cells?

13. Through what structure does oxygen from the mother pass into the fetus's body?

14. What are four stages of human life following birth?

15. What two cells combine to make a zygote?

16. What is the difference between budding and fragmentation?

Concept Mapping

17. Use the following terms to create a concept map: asexual reproduction, budding, external fertilization, fragmentation, reproduction, internal fertilization, sexual reproduction.

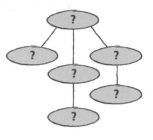

CRITICAL THINKING AND PROBLEM SOLVING

Write one or two sentences to answer the following questions:

18. Explain why the testes are found in the scrotum instead of inside the male body.

19. What is the function of the uterus? How is its function related to the menstrual cycle?

20. How is meiosis important to human reproduction?

MATH IN SCIENCE

21. Hardy Junior High School has 2,750 students. If 1 pair of identical twins is born for every 250 births, about how many pairs of identical twins will be attending the school?

22. Mrs. Schmidt had a baby April 30th. Her baby developed inside her uterus for 9 months. What month was her egg fertilized?

23. In the United States, seven infants die before their first birthday for every 1,000 births. Convert this figure to a percentage. Is your answer greater than or less than 1 percent?

24. In Haiti, a small country in the Caribbean, 74 infants die before their first birthday for every 1,000 births. Convert this figure to a percentage. Is your answer greater or less than 1 percent? Why do you think there is such a difference between the United States and Haiti?

INTERPRETING GRAPHICS

The following graph illustrates the cycles of the male hormone, testosterone, and the female hormone, estrogen. The blue line shows the estrogen level in a female over a period of 28 days. The red line shows the testosterone level in a male over a period of 28 days.

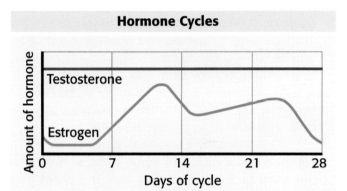

Hormone Cycles

25. What is the major difference between the two hormone levels over the 28-day period?

26. What cycle do you think estrogen affects?

27. Why might the level of testosterone stay the same?

Reading Check-up

Take a minute to review your answers to the Pre-Reading Questions found at the bottom of page 100. Have your answers changed? If necessary, revise your answers based on what you have learned since you began this chapter.

Acne

If you are a teenager, you probably have some firsthand experience with acne. If you don't, you probably will. And contrary to what you may have heard, acne is not caused by greasy foods and candy, though these foods can aggravate the problem. The hormonal fluctuations that occur as young people mature into adults often cause acne.

What Are Pimples?

Skin contains thousands of tiny pores. Each pore contains sebaceous (suh BAY shuhs) glands that produce sebum, the oil you may have noticed on the surface of your skin. This oil is necessary to maintain healthy skin. The production and release of sebum is stimulated by androgens, the male sex hormones, which become active in both girls and boys during puberty.

Sebum usually escapes from the pores without a problem. But sometimes skin cells do not shed properly, and they clog the pores. The sebum that collects in the pores causes lesions, commonly called pimples.

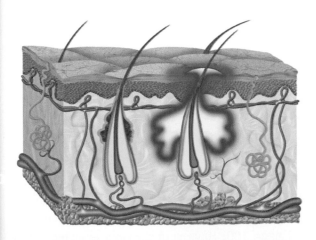

▲ *Acne is caused by the buildup of sebum and dead cells in the pores of the skin.*

Learn Your Lesions

There are two kinds of lesions—noninflamed lesions and inflamed lesions. Noninflamed lesions include blackheads and whiteheads. Some people think blackheads are pores filled with dirt. The dark color of these lesions is actually the result of dark skin pigments or oil trapped in the pores. Whiteheads are white because their contents are hidden under the skin's surface. Inflamed lesions are caused by bacteria and are often red and swollen. Bacteria live in healthy pores, and when pores become clogged, the bacteria are trapped and can cause irritation and infection.

Heredity

Family history appears to be a factor in the development of acne. Unfortunately, if your parents or brothers and sisters had acne, you are likely to have acne too. The causes of hereditary acne remain unclear. Your skin may be genetically programmed to produce more sebum than is produced in other teenagers.

Is There Hope?

Certain over-the-counter products can clean the dead skin cells and sebum out of the pores. Many medications inhibit the production of sebum or encourage the shedding of skin cells. Sometimes doctors prescribe antibiotics, such as tetracycline or erythromycin, to treat severe cases of acne. Most acne clears up as people become adults.

On Your Own

▶ Find out what the active ingredient is in an over-the-counter acne medication. Do some research on this ingredient to find out how it works. Report your findings to the class.

Technology in Its Infant Stages

Every year thousands of babies are born with life-threatening diseases or severe birth defects. What if medical treatments were available to these babies before they were born? Doctors at San Francisco, Harvard, and Vanderbilt Universities are performing experimental fetal surgery with encouraging results.

When Is Fetal Surgery an Option?

To date, approximately 100 fetal operations have been performed across the country. Corrective treatments can take place between the 18th and 30th weeks of pregnancy. Many factors determine whether fetal surgery is appropriate. Surgery is considered to be an option only if the condition is life threatening. However, fetuses with several defects or chromosomal abnormalities are not eligible for surgery.

Successful surgeries have been performed on fetal patients with spina bifida, diaphragmatic hernias, malformations of the lungs, and urinary tract obstructions. Spina bifida is a defect that leaves the spine exposed. A diaphragmatic hernia is a hole in the diaphragm. This condition causes severe breathing difficulties.

Surgery on a Small Scale

Fetal surgery can fall into one of three categories. The least traumatic type of treatment uses a laser scalpel or an endoscope. The scalpel is used to remove chest tumors. An *endoscope* is a video-guided tool that combines a camera lens and scissors that are less than 0.2 cm wide. The doctor guides the scissors through a tiny cut in the abdominal and uterine walls. The doctor is unable to see the fetus directly during this surgery because the cut is so small. Therefore, he or she must watch the video images provided by the endoscope during the operation.

A more traumatic option is open fetal surgery. In this treatment, the mother's abdomen and uterus are opened, and the fetus is partially exposed.

The third, and relatively new, option is called fetal stem cell transplant. This treatment is essentially a bone marrow transplant for the fetus. It is used to treat genetic diseases and diseases of the immune system.

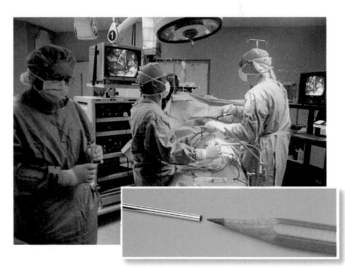

▲ *The endoscope shown here is used to perform fetal surgery.*

What the Future Holds

Each fetal surgery results in the improvement of techniques and treatments, as well as in the expansion of the types of defects and diseases that can be treated. As the number of fetal surgeries increases, fetal surgery will become much more routine.

Going Further

▶ The endoscopes used in fetal surgery use a technology called fiber optics. Research what items around your home also use fiber optics.

123

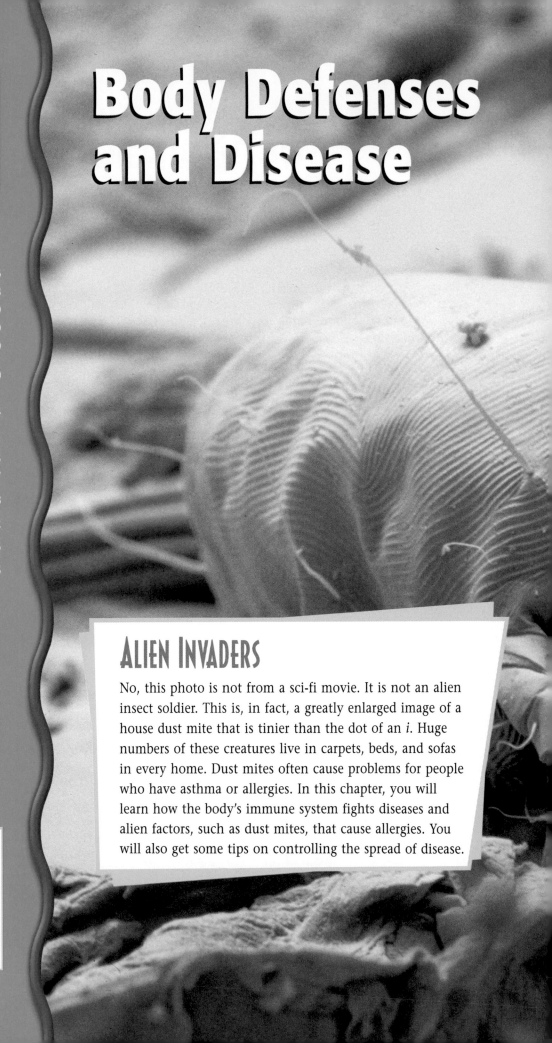

CHAPTER

6

Body Defenses and Disease

Pre-Reading Questions

1. When you "catch a cold," what is it that infects your body?

2. How does a fever help you get well?

ALIEN INVADERS

No, this photo is not from a sci-fi movie. It is not an alien insect soldier. This is, in fact, a greatly enlarged image of a house dust mite that is tinier than the dot of an *i*. Huge numbers of these creatures live in carpets, beds, and sofas in every home. Dust mites often cause problems for people who have asthma or allergies. In this chapter, you will learn how the body's immune system fights diseases and alien factors, such as dust mites, that cause allergies. You will also get some tips on controlling the spread of disease.

INVISIBLE INVADERS

In this activity, you will use a technique that makes "invisible" lifeforms become visible.

Procedure

1. Obtain two **Petri dishes containing nutrient agar.** Label them "Washed" and "Unwashed."

2. Rub two **marbles** between the palms of your hands. Observe the appearance of the marbles.

3. Roll one marble in the Petri dish labeled "Unwashed."

4. Put on a pair of **disposable gloves.** Wash the other marble with **soap** and **warm water** for 4 minutes. Does the appearance of the marble change after it is washed?

5. Roll the washed marble in the Petri dish labeled "Washed."

6. Secure the lids of the Petri dishes with **transparent tape**. Place the dishes in a warm, dark place. **Caution:** Do not open the Petri dishes after they are sealed.

7. Observe the Petri dishes each day for a week. Record your observations in your ScienceLog.

Analysis

8. How did the washed and unwashed marbles compare? How did the Petri dishes differ after several days?

9. Why is it important to wash your hands before eating?

What You'll Do

♦ Explain the difference between infectious diseases and non-infectious diseases.

♦ Identify five ways that you might come into contact with a pathogen.

♦ Discuss four methods that have helped reduce the spread of disease.

Disease

You've probably heard it before: "Cover your mouth when you sneeze!" "Wash your hands!" "Don't put that in your mouth!" What is all the fuss about? When people say these things to you, they are concerned about the spread of disease.

What Causes Disease?

When you have a *disease,* your normal body functions are disrupted. Some diseases, such as most cancers and heart disease, are not spread from one person to another. They are called **noninfectious diseases.**

Noninfectious diseases can be caused by a variety of factors. For example, a genetic disorder causes the disease hemophilia, in which a person's blood does not clot properly. The disease scurvy is caused by a lack of vitamin C in the diet. Smoking, lack of physical activity, and a high-fat diet can greatly increase a person's chances of getting certain noninfectious diseases. Avoiding harmful habits may help you avoid noninfectious diseases.

A disease that can be passed from one living thing to another is an **infectious disease.** Infectious diseases are caused by agents called **pathogens.** Viruses and some bacteria, fungi, protists, and worms may all cause diseases. **Figure 1** shows some common pathogens.

Figure 1 *Pathogens, such as these, are often referred to as germs.*

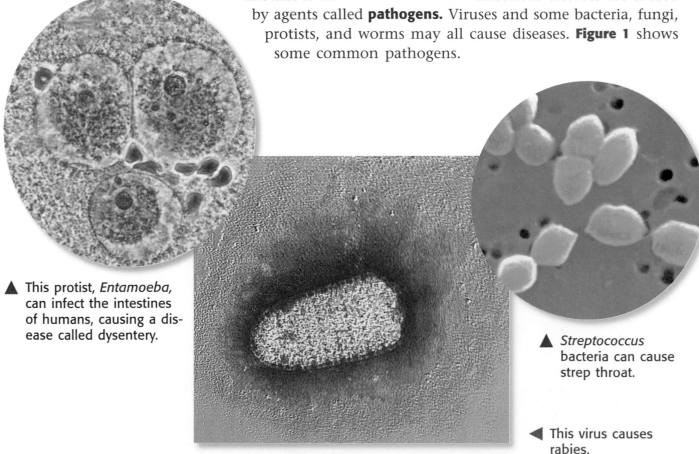

▲ This protist, *Entamoeba,* can infect the intestines of humans, causing a disease called dysentery.

▲ *Streptococcus* bacteria can cause strep throat.

◀ This virus causes rabies.

Pathways to Pathogens

There are many ways pathogens can be passed from one person to another. Being aware of them can help you stay healthy.

Through the Air Some pathogens travel through the air. For example, a single sneeze, like the one shown in **Figure 2,** releases thousands of tiny droplets of moisture that can carry pathogens.

Contaminated Objects A person who is sick may leave bacteria or viruses on objects such as doorknobs, keyboards, drinking glasses, towels, or combs. If you drink from a glass that an infected person has just used, you could become infected with a pathogen.

Person to Person Some pathogens are spread by direct person-to-person contact. You can become infected with some illnesses by kissing, shaking hands, or touching the sores of an infected person.

Animals Some pathogens are carried by animals. For example, humans can get a fungus called ringworm from handling an infected dog or cat. Also, ticks may carry bacteria that cause Lyme disease or Rocky Mountain spotted fever.

Food and Water Drinking water in the United States is generally safe, but water lines can break, or treatment plants can become flooded, allowing microorganisms to enter the public water supply. Bacteria growing in foods and beverages can cause illness too. Refrigerating foods can slow the growth of many of these pathogens, but meat, fish, and eggs that are not cooked enough can still contain dangerous bacteria or parasites. Leaving food out at room temperature can give bacteria such as *Salmonella* time to grow and produce toxins in the food. For these reasons, it is important to wash all used cooking tools.

Figure 2 *A sneeze can force thousands of pathogen-carrying droplets out of your body at up to 160 km per hour.*

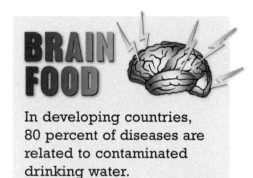

In developing countries, 80 percent of diseases are related to contaminated drinking water.

Self-Check

Jackie cut up raw meat on her kitchen counter. If her brother makes a sandwich on the same counter later, how could he come in contact with a pathogen? *(See page 212 to check your answer.)*

Physics

C O N N E C T I O N

Hospitals use a machine called an autoclave to kill bacteria on surgical instruments. An autoclave works by increasing the pressure of steam as its temperature increases. The combined effect of pressure and temperature kills bacteria at a lower temperature than would normally be needed.

Putting Pathogens in Their Place

Until the twentieth century, surgery patients often died of bacterial infections. But as doctors learned more about disease, it became clear that simple cleanliness could help prevent the spread of some diseases. Today, hospitals and clinics use a variety of technologies to prevent the spread of pathogens. For example, ultraviolet radiation, boiling water, and chemicals are used in health facilities to kill pathogens.

Pasteurization During the mid-1800s, Louis Pasteur, a French scientist, discovered that microorganisms caused wine to spoil. The uninvited microorganisms were bacteria. Pasteur devised a method of using heat to kill most of the bacteria in the wine. This method is called *pasteurization,* and it is still used today. The milk that the girl in **Figure 3** is drinking has been pasteurized.

Vaccines and Immunity In the late 1700s, no one knew what a pathogen was. During this time, British physician Edward Jenner studied a disease called smallpox. He observed that people who had been infected with cowpox seemed to have protection against smallpox. This protection, or resistance to a disease, is called **immunity.** Jenner's work led to the first modern *vaccine.* A vaccine is a substance that helps your body develop immunity to a disease.

Today vaccines are used all over the world to prevent many serious diseases. Modern vaccines contain pathogens that are killed or specially treated so that they can't make you very sick. The vaccine is enough like the pathogen to allow your body to develop a defense against the disease.

Figure 3 *Today pasteurization is used to kill pathogens in many different types of food, including dairy products, eggs, meats, and juices.*

Antibiotics Bacterial infections can be a serious threat to your health. Fortunately, doctors can usually treat these kinds of infections with antibiotics. An *antibiotic* is a substance that can kill bacteria or slow the growth of bacteria. Antibiotics may also be used to treat infections caused by other microorganisms, like fungi. If you take an antibiotic when you are sick, it is important that you take it according to your doctor's instructions to ensure that all the pathogens are killed.

Viruses, such as those that cause colds, are not affected by antibiotics. The only way to destroy viruses in your body is to locate and kill the cells they have invaded. In the next section, you'll see how a healthy immune system does just that.

Cold Calamity

Frank caught a bad cold just before the opening night of his school play. He visited his doctor and asked her to prescribe antibiotics for his cold. The doctor politely refused and suggested that Frank stay home and get plenty of rest. Why do you think the doctor refused to give Frank antibiotics? Explain your answer.

SECTION REVIEW

1. How is an infectious disease different from a noninfectious disease?

2. List five ways that you might come into contact with a pathogen.

3. How does a vaccine work?

4. **Inferring Relationships** Why might the risk of infectious disease be high in a community that has no water-treatment facility?

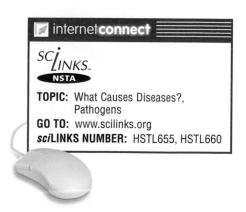

SECTION **2**
READING WARM-UP

Terms to Learn

immune system   B cell
macrophage   antibody
T cell   memory B cell

What You'll Do

◆ Describe how your body keeps out pathogens.
◆ Explain how the immune system works.
◆ Discuss the purpose of a fever.

Your Body's Defenses

Although you probably don't realize it, your body must constantly protect itself against pathogens that are trying to invade it. But how does your body do that? Luckily, your body has its own built-in defense system.

Your Suit of Armor

For a pathogen to harm you, it must attack a part of your body. Usually, though, only a small percentage of the pathogens around you ever make it past your first lines of defense.

Eyes, Nose, and Mouth Many organisms that try to enter your eyes or mouth are destroyed by special enzymes. Pathogens that enter your nose are washed down the back of your throat by mucus. The mucus carries the pathogens to your stomach, where most are quickly digested.

Skin Your skin is made of many layers of flat cells. The outermost layers are dead. As a result, any pathogen that lands on your skin cannot find a live cell to infect. As **Figure 4** shows, the dead skin cells are constantly dropping off of your body as new skin cells grow from beneath. As the dead skin cells flake off, they carry away viruses, bacteria, and other microorganisms. In addition, glands secrete oil onto your skin's surface. The oil contains chemicals that kill many pathogens.

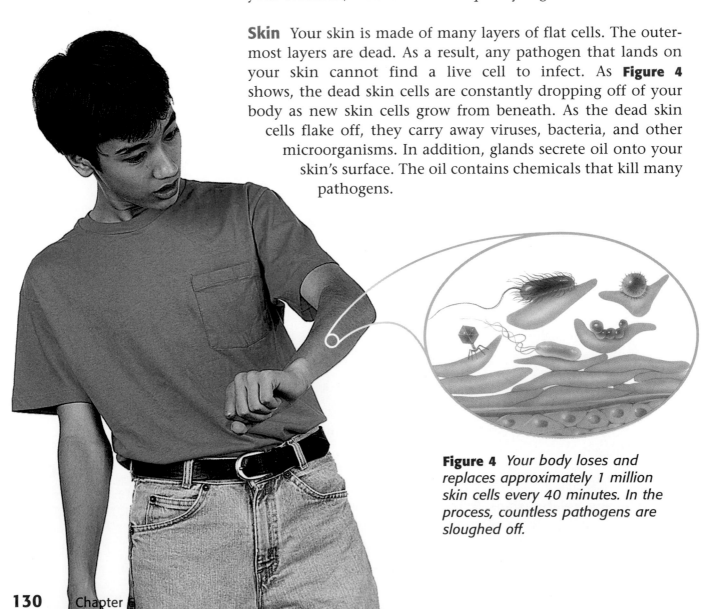

Figure 4 *Your body loses and replaces approximately 1 million skin cells every 40 minutes. In the process, countless pathogens are sloughed off.*

130   Chapter 6

A Forced Entry

Sometimes skin is burned, cut, or punctured. When this happens, pathogens can enter the body. The body acts quickly to keep out as many pathogens as possible. Blood flow to the injured area increases. Cell parts in the blood called *platelets* help seal the open wound so that no more pathogens can enter.

The increased blood flow also brings cells that belong to the **immune system,** the body system that fights pathogens. The immune system is not localized in any one place in your body, nor is it controlled by any one organ, such as the brain. Instead, it is an army of individual cells, tissues, and organs that work together to combat invading pathogens.

Soldiers of the Immune System

The immune system consists mainly of three kinds of cells. One kind is the **macrophage** (MAK roh FAYJ). Macrophages engulf, or eat, any microorganisms or viruses floating around. If only a few microorganisms and viruses have entered the wound, the macrophages can easily stop them.

The other two main types of immune-system cells are **T cells** and **B cells.** T cells play an important role in coordinating the immune system. Many B cells make **antibodies,** which are proteins that attach to specific pathogens. Your body is capable of making billions of different antibodies, but each antibody usually attaches to only one type of pathogen, as illustrated in **Figure 5.**

It's Only Skin Deep

Cut an **apple** in half. Place **plastic wrap** over both halves. The plastic wrap will act as skin. Use **scissors** to cut the plastic wrap on one of the apple halves, then use an **eye-dropper** to drip **food coloring** on each apple half. The food coloring represents pathogens coming into contact with your body. Now, answer the following questions:

1. What happened to each apple half?

2. How is the plastic wrap similar to skin?

3. How is the plastic wrap different from skin?

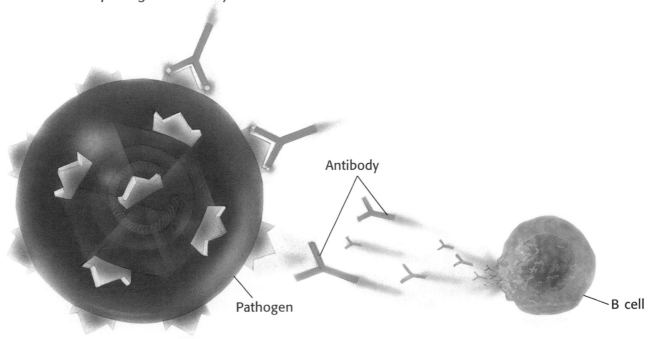

Figure 5 *An antibody's shape is very specialized. It matches a pathogen like a key fits a lock.*

Antibody

Pathogen

B cell

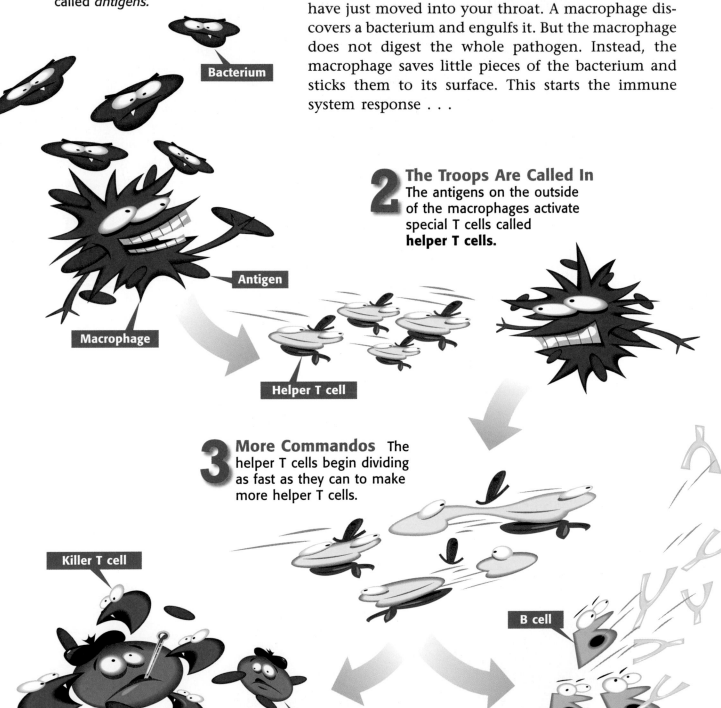

1 **Enemy Invasion!** A macrophage displays the pieces of bacteria it has engulfed for other cells to notice. The pieces of bacteria are called *antigens*.

Bacterium

Antigen

Macrophage

Helper T cell

A War Within

Macrophages can take care of a small invasion of pathogens, but what would happen if millions of pathogens invaded your body? Imagine that bacteria have just moved into your throat. A macrophage discovers a bacterium and engulfs it. But the macrophage does not digest the whole pathogen. Instead, the macrophage saves little pieces of the bacterium and sticks them to its surface. This starts the immune system response . . .

2 **The Troops Are Called In** The antigens on the outside of the macrophages activate special T cells called **helper T cells.**

3 **More Commandos** The helper T cells begin dividing as fast as they can to make more helper T cells.

Killer T cell

B cell

Infected cell

4 **Assassins** The helper T cells send word to another kind of T cell called a **killer T cell.** Killer T cells kill any cell infected with pathogens. They recognize an infected cell because the cell places antigens on its surface.

5 **Foot Soldiers** The helper T cells also activate **B cells.** The B cells then make millions of antibodies.

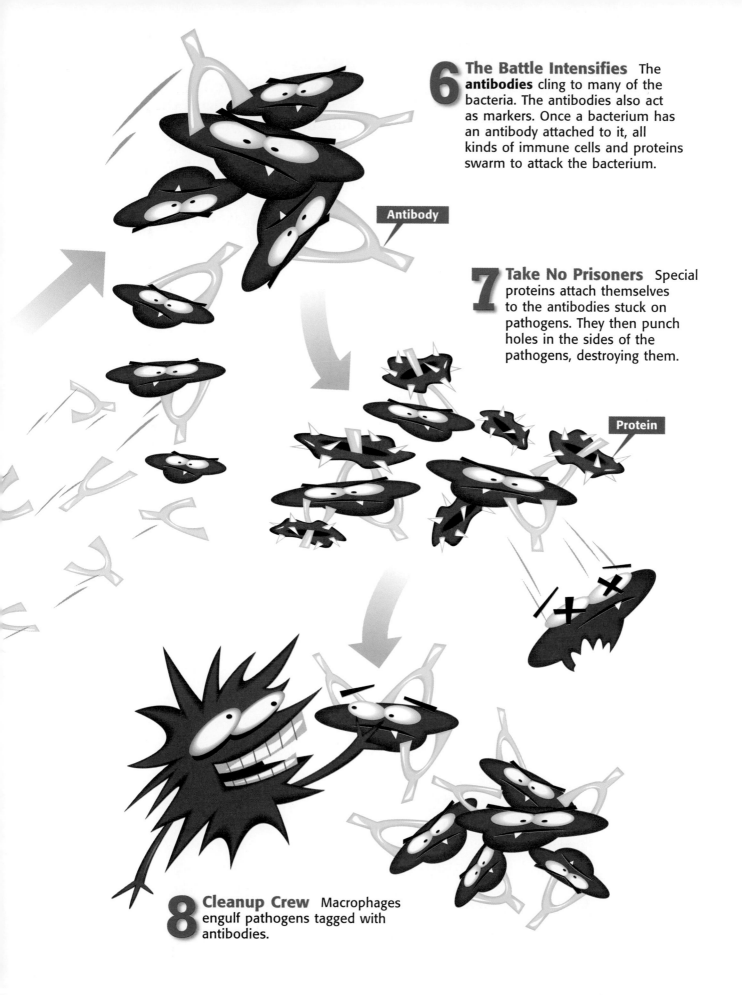

6 The Battle Intensifies The **antibodies** cling to many of the bacteria. The antibodies also act as markers. Once a bacterium has an antibody attached to it, all kinds of immune cells and proteins swarm to attack the bacterium.

Antibody

7 Take No Prisoners Special proteins attach themselves to the antibodies stuck on pathogens. They then punch holes in the sides of the pathogens, destroying them.

Protein

8 Cleanup Crew Macrophages engulf pathogens tagged with antibodies.

Heating Things Up

When macrophages activate the helper T cells, they also send a chemical signal that tells your brain to turn up the thermostat. In a few minutes, your body's temperature can rise several degrees. A moderate fever of one or two degrees actually helps you get well faster because it slows the growth of some pathogens. As is shown in **Figure 6,** a fever also helps B cells and T cells multiply faster than usual.

Haven't We Met Somewhere Before?

The immune system responds very quickly if your B cells recognize the invading pathogen and can produce antibodies for it. However, B cells must have had previous contact with a pathogen before they can make the correct antibodies. During the first encounter with a new pathogen, specialized B cells make antibodies that are effective against that particular invader. This process takes about 2 weeks, which is far too long to prevent an infection. Therefore, the first time you are infected, you usually get sick.

A few of the B cells become **memory B cells** that "remember" how to make an antibody for a particular pathogen. If the pathogen shows up again, the memory B cells produce B cells that make enough antibodies to protect you in just 3 or 4 days.

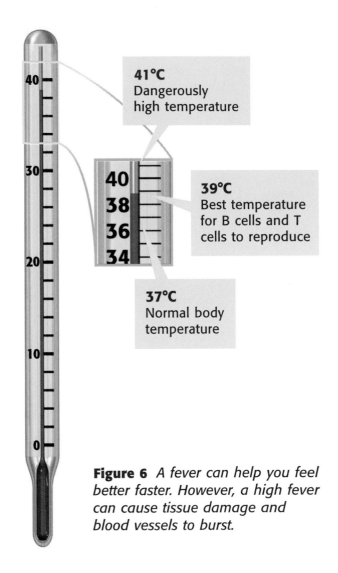

41°C
Dangerously high temperature

39°C
Best temperature for B cells and T cells to reproduce

37°C
Normal body temperature

40
38
36
34

Figure 6 *A fever can help you feel better faster. However, a high fever can cause tissue damage and blood vessels to burst.*

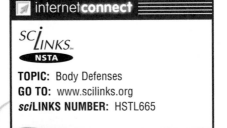
SECTION REVIEW

1. List three ways your body defends itself against pathogens.

2. Name three different cells in the immune system, and describe how they respond to pathogens.

3. Can you make antibodies for diseases you have never come in contact with? Why or why not?

4. **Applying Concepts** If you had chickenpox at age 7, what would be your chances of getting chickenpox again if your memory B cells lived only 2 months?

Terms to Learn

allergy   cancer
autoimmune
  disease

What You'll Do

◆ Explain the difference between allergies and autoimmune diseases.

◆ Discuss what cancer is.

◆ Describe how HIV affects the immune system.

Challenges to the Immune System

The immune system is a very effective body-defense system, but it is not invincible. There are some diseases that the immune system is unable to deal with. There are also conditions in which the immune system does not work properly.

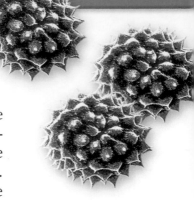

Ragweed pollen

A-a-achoo!

Sometimes the immune system overreacts to antigens that are not dangerous to the body. This inappropriate reaction is called an **allergy.** Allergies may be caused by many things, including certain foods and medicines. Some of the culprits behind allergic reactions are shown in **Figure 7.** Symptoms can range from a runny nose and itchy eyes to more serious conditions, such as asthma.

Doctors are not sure why the immune system overreacts in some people. Scientists think allergies might be useful because the mucus draining from your nose carries away pollen, dust, and microorganisms.

Figure 7 Things That Cause Allergies

Pollen

Dust mite

Animal hair and dander (skin flakes)

Cigarette smoke

Body Defenses and Disease   **135**

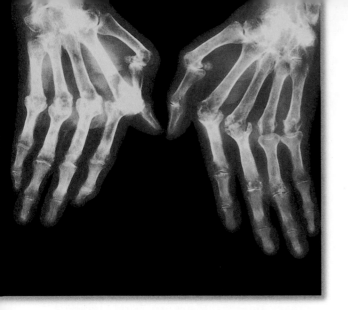

Figure 8 *In rheumatoid arthritis, immune-system cells cause joint-tissue swelling, which can lead to joint deformities.*

Autoimmune Diseases

An **autoimmune disease** is a disease in which the immune system attacks the body's own cells. This happens when immune-system cells are not able to tell the difference between pathogens and particular body cells. One autoimmune disease is rheumatoid arthritis, in which the immune system attacks the joints. The most common location for rheumatoid arthritis is the joints of the hands, as shown in **Figure 8.** Other autoimmune diseases include type 1 diabetes, Graves' disease, multiple sclerosis, and lupus.

Cancer

Healthy cells divide at a carefully regulated rate. Occasionally, though, a cell doesn't respond to the body's regulation and begins dividing at an uncontrolled rate. As can be seen in **Figure 9,** killer T cells destroy this type of cell. But sometimes division of these cells gets out of the control of the immune system. This causes a condition known as **cancer.**

Many cancers will invade nearby tissues. They can also enter the cardiovascular system or lymphatic system. This way, cancers can be transported to other places in the body. Cancers disrupt the normal activities of organs they have invaded, often leading to death. Today, though, there are many treatments for cancer. Radiation and certain drugs can be used to kill cancer cells or slow their division.

Figure 9 The Destruction of an Unregulated Cell

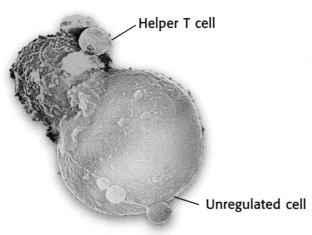

Helper T cell

Unregulated cell

❶ A killer T cell attacks an unregulated cell.

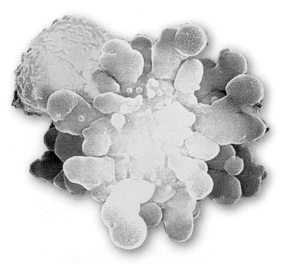

❷ The cell's membrane ruptures as the cell dies.

AIDS

The human immunodeficiency virus (HIV) causes the acquired immune deficiency syndrome (AIDS). Most viruses infect cells in the nose, mouth, lungs, or intestines, but HIV is different. As you can see in **Figure 10,** HIV infects the immune system itself, using helper T cells as factories to produce more viruses. The helper T cells are destroyed in the process. Remember that the helper T cells put the B cells and killer T cells to work.

People with AIDS have very few helper T cells, so nothing activates the B cells and killer T cells. Therefore, the immune system cannot attack HIV or any other pathogen. People with AIDS don't usually die of AIDS itself. They die of other diseases that they are unable to fight off.

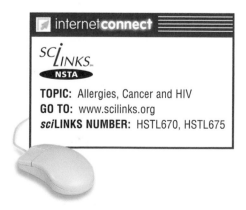

Figure 10 *The blue particles on this helper T cell are human immunodeficiency viruses. They were made inside the cell and can now go and infect other cells.*

SECTION REVIEW

1. What is the difference between allergies and autoimmune diseases?

2. Why is it important for immune-system cells to be able to recognize all of the body's own cells?

3. What characterizes a cancerous cell?

4. **Interpreting Graphs** Over time, people with AIDS become very sick and are unable to fight off infection. Use the information in the graph below to explain why this occurs.

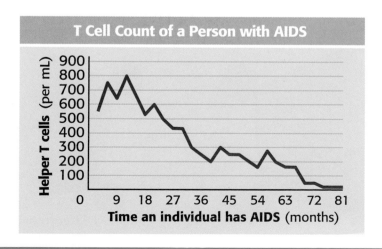

T Cell Count of a Person with AIDS

Helper T cells (per mL)

Time an individual has AIDS (months)

internet**connect**

SC*i*LINKS.
NSTA

TOPIC: Allergies, Cancer and HIV
GO TO: www.scilinks.org
*sci*LINKS NUMBER: HSTL670, HSTL675

Making Models Lab

Antibodies to the Rescue

Some cells of the immune system, called B cells, make antibodies that attack and kill invading viruses and microorganisms. These antibodies help make your body immune to disease. Have you ever had chickenpox? If you have, your body has built up antibodies that can recognize that particular virus. Antibodies will attach themselves to the virus, tagging it for destruction. If you are exposed to the same disease again, the antibodies remember that virus. They will attack the virus even quicker and in greater number than they did the first time. That is why you will probably never have chickenpox more than once.

In this activity, you will construct simple models of viruses and their antibodies. You will see how antibodies are specific for a particular virus.

MATERIALS

- craft materials, such as buttons, fabric scraps, pipecleaners, and recycled materials
- scissors
- tape or glue
- colored paper

Procedure

1 Draw the virus patterns shown on this page on a separate piece of paper, or design your own virus models from the craft supplies. Remember to design different receptors on each of your virus models.

2 In your ScienceLog, write a few sentences describing how your viruses are different.

3 Cut out the viruses, and attach them to a piece of colored paper with tape or glue.

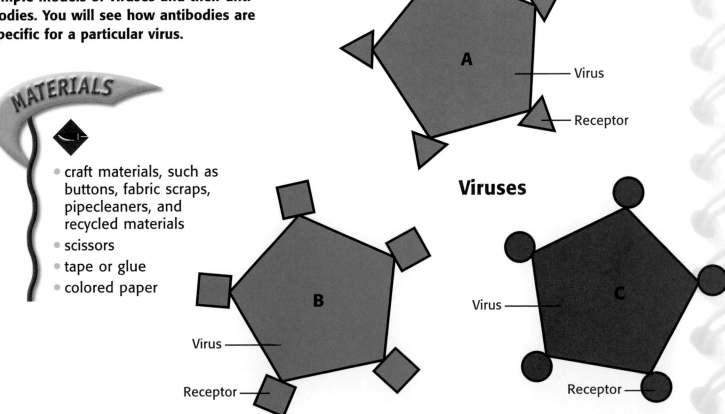

Viruses

Antibodies

4 Select the antibodies drawn above or design your own antibodies that will exactly fit on the receptors on your virus models. Draw or create each antibody enough times to attach one to each receptor site on the virus.

5 Cut out the antibodies you have drawn. Arrange the antibodies so that they bind to the virus at the appropriate receptor. Attach them to the virus with tape or glue.

Analysis

6 Explain how an antibody "recognizes" a particular virus.

7 After the attachment of antibodies to the receptors, what would be the next step in the immune response?

8 Many vaccines use weakened copies of the virus to protect the body. Use the model of a virus and its specific antibody to explain how vaccines work.

9 Use your model of a virus to demonstrate to the class how a receptor might change or mutate so that a vaccine would no longer be effective.

Going Further

Research in the library or on the Internet to find information about the discovery of the Salk vaccine for polio. Include information on how polio affects people today.

Research in the library or on the Internet to find information about filoviruses. What do they look like? What diseases do they cause? Why are they especially dangerous? Is there an effective vaccine against any filovirus? Explain.

Chapter Highlights

SECTION 1

Vocabulary

noninfectious disease *(p. 126)*

infectious disease *(p. 126)*

pathogen *(p. 126)*

immunity *(p. 128)*

Section Notes

- A disease disrupts the body's ability to function normally.

- Noninfectious diseases cannot be spread from one person to another.

- Infectious diseases are caused by pathogens that are passed from one living thing to another.

- Pathogens are agents such as viruses or microorganisms that can make us sick.

- Pathogens can travel through the air or can be spread by contact with other people, contaminated objects, animals, or food.

- Cleanliness, pasteurization, vaccines, and antibiotics help control the spread of pathogens.

Labs

Passing the Cold *(p. 184)*

SECTION 2

Vocabulary

immune system *(p. 131)*

macrophage *(p. 131)*

T cell *(p. 131)*

B cell *(p. 131)*

antibody *(p. 131)*

memory B cell *(p. 134)*

Section Notes

- Enzymes in your eyes, nose, and mouth kill most pathogens that try to enter. Other pathogens are washed down the throat and destroyed in the stomach.

- Dead skin cells and oil help to keep germs out of the body.

☑ Skills Check

Math Concepts

SPREAD OF DISEASES It is easy to infect a large group of people with a disease. For instance, suppose a man with the flu gets onto an empty train. The train stops at 10 different towns. If five people get off and five people get on at every stop, how many people could the man expose to his illness?

10 stops × 5 people = 50 people exposed

Visual Understanding

IMMUNE RESPONSE
Look at the immune response illustration on pp. 132–133. Review each step to make sure you understand how your immune system works. Think about how many different cells the immune system uses to destroy pathogens. Also, notice that each cell has a special job.

SECTION 2

- When pathogens get into your blood or tissues, the immune system reacts.

- Macrophages engulf pathogens. Macrophages then display parts of the pathogens, called antigens, on their surface.

- Macrophages activate helper T cells. The helper T cells put the killer T cells and B cells to work.

- Killer T cells kill infected cells. B cells make antibodies.

- Antibodies cling to antigens and attract macrophages and other cells. Special proteins kill pathogens with antibodies stuck to them.

- Macrophages cause fever, which speeds the division of T cells and B cells.

- Memory B cells stand ready to produce more B cells that make antibodies if the pathogen appears again.

SECTION 3

Vocabulary

allergy *(p. 135)*
autoimmune disease *(p. 136)*
cancer *(p. 136)*

Section Notes

- The immune system can overreact to a harmless antigen. This reaction is called an allergy.

- Autoimmune diseases are diseases in which the immune system attacks the body's healthy tissue.

- Cancer cells can enter the body's circulatory systems and infect other areas of the body.

- HIV attacks helper T cells, preventing the immune system from functioning properly.

Chapter Review

USING VOCABULARY

To complete the following sentences, choose the correct term from each pair of terms listed below:

1. Diseases caused by pathogens are __?__. (*infectious* or *noninfectious*)

2. Antibiotics can be used to kill some __?__. (*bacteria* or *viruses*)

3. Macrophages attract __?__. (*helper T cells* or *killer T cells*)

4. Certain B cells make __?__. (*antigens* or *antibodies*)

5. An immune system overreaction to a harmless substance is a(n) __?__. (*allergy* or *vaccine*)

6. __?__ attacks helper T cells. (*HIV* or *Cancer*)

UNDERSTANDING CONCEPTS

Multiple Choice

7. Pathogens are
   a. all viruses and microorganisms.
   b. viruses and microorganisms that cause disease.
   c. noninfectious organisms.
   d. all bacteria that live in water.

8. The following is an infectious disease:
   a. allergies
   b. rheumatoid arthritis
   c. asthma
   d. common cold

9. The skin keeps pathogens out by
   a. staying warm enough to kill pathogens.
   b. releasing killer T cells onto the surface.
   c. shedding dead cells and secreting oils.
   d. All of the above

10. Memory B cells
    a. kill pathogens.
    b. activate killer T cells.
    c. activate killer B cells.
    d. produce B cells that make antibodies.

11. A fever
    a. slows pathogen growth.
    b. helps B cells multiply faster.
    c. helps T cells multiply faster.
    d. All of the above

12. Macrophages
    a. make antibodies.
    b. release helper T cells.
    c. live in the gut.
    d. engulf pathogens.

Short Answer

13. Explain how macrophages start an immune response.

14. Describe the role of helper T cells in responding to an infection.

Concept Mapping

15. Use the following terms to create a concept map: macrophages, helper T cells, B cells, antibodies, antigens, killer T cells, memory B cells.

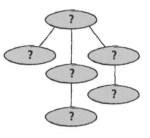

CRITICAL THINKING AND PROBLEM SOLVING

Write one or two sentences to answer the following questions:

16. Why does the disappearance of helper T cells in AIDS patients damage the immune system?

17. Many people take fever-reducing drugs as soon as their temperature exceeds 37°C. Why might it not be a good idea to immediately reduce a fever with drugs? What are the benefits of taking fever-reducing drugs?

18. The risk of dying from a whooping cough vaccine is about one in 1 million. In contrast, the risk of dying from whooping cough itself is about one in 500. Discuss the pros and cons of this vaccination.

MATH IN SCIENCE

19. Suppose you have 50,000 flu viruses on your fingers and you rub your eyes. Only 20,000 viruses make it into your eyes, 10,000 dissolve in chemicals, and 10,000 are washed down into your nose. Of those, you sneeze out 2,000. How many viruses are left to wash down the back of your throat and start an infection?

INTERPRETING GRAPHICS

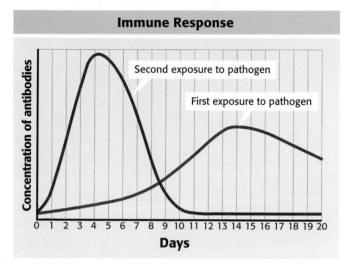

Immune Response

The graph above compares the concentration of antibodies in the blood the first time you are exposed to a pathogen with the concentration of antibodies the next time you are exposed to the pathogen.

20. Are there more antibodies present during the first week of the first exposure or the first week of the second exposure? Why do you think this is so?

21. What is the difference in recovery time between the first exposure and second exposure? Why?

Reading Check-up

Take a minute to review your answers to the Pre-Reading Questions found at the bottom of page 124. Have your answers changed? If necessary, revise your answers based on what you have learned since you began this chapter.

CAREERS

NATUROPATHIC PHYSICIAN

Dr. Stacey Kargman of Tucson, Arizona, is a doctor of naturopathic medicine (NMD), commonly referred to as a naturopath. An NMD has similar training to an MD but is less likely than a traditionally trained doctor to use prescription drugs or surgery to treat a patient's symptoms. Naturopaths tend to look for a natural way to treat a patient, using drugs or surgery as a last resort. Dr. Kargman tries to strengthen her patients' immune systems by focusing on things like nutrition.

On Your Own

▶ Do some research about naturopaths. Find out how an NMD's training and practice differ from the training and practice of an MD.

Dr. Kargman attended the Southwest College of Naturopathic Medicine, where she studied all the sciences a medical doctor would study—like biochemistry, anatomy, pharmacology, and physiology. Beyond the standard medical school sciences, naturopaths spend an additional four years studying subjects like botanical medicines, homeopathy, acupuncture, counseling, and nutrition. "Naturopathy is a way of looking at the person as a whole," says Kargman.

The Keystone to Good Health

Many naturopaths believe that nutrition is the keystone to good health. "Most MDs don't talk to their patients about their diets," Kargman explains. "I'm in a position to talk to them about what they eat and how it may be affecting their health. Food allergies can cause an immune reaction in the body—anything from depression to skin problems to migraine headaches. Even though I can prescribe prescription medications, I usually defer to MDs when it comes to prescription medications."

Dr. Kargman treats many HIV and AIDS patients. She encourages these patients and others who need prescription medications to work with their medical doctor and their naturopath at the same time. That way, patients get the best care.

A Fulfilling Career

Dr. Kargman says the best part of her work is making people feel better. "Someone might come to me and say they have terrible migraines that they can no longer live with and that they've seen every doctor. After examining them, I might be able to tell them something as simple as, 'Stop eating wheat.' The simplest thing can change someone's life . . . It's not like putting a bandage on it. It's fixing the cause of the problem."

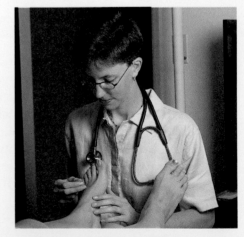

▲ *Stacey Kargman, NMD, tries to treat the patient as a whole.*

Frogs in the Medicine Cabinet?

Frog skin, mouse intestines, cow lungs, and shark stomachs—sounds like the ingredients for a witch's brew, doesn't it? Actually, these animal parts are being tested in an effort to create more effective medicines to combat harmful bacteria.

Leapin' Lily Pads—It's Infection Protection

In 1896, a biologist named Michael Zasloff was studying African clawed frogs. He noticed that cuts in the frogs' skin healed quickly and never became infected. Zasloff decided to investigate further. He found that when a frog was cut, its skin released a liquid antibiotic that killed invading bacteria.

Scientists have found other animals whose bodies contain similar infection fighters. For example, the stomach and tissues of sand sharks (also called dogfish) contain chemicals that kill bacteria and other microorganisms. These useful antibiotics are also in moths, pigs, mice, cows, and even the small intestines of humans!

What's in Dog Spit?

A healthy dog licks cuts, scrapes, and minor wounds to clean them. A mother cat licks her kittens clean. Have you ever wondered why animals do that? Well, dogs, cats, humans, and some other animals have an antibacterial enzyme in their saliva. When animals lick a wound, the enzymes kill the bacteria and help the wound heal.

▲ *African clawed frogs produce a natural antibiotic.*

POW! Punching Holes in Bacteria

Bacteria are becoming resistant to many man-made antibiotics, which means that the drugs no longer affect the bacteria. Scientists now face the challenge of developing new antibiotics that can overcome the resistant strains of bacteria.

Antibiotics from animals pack a different punch than some man-made antibiotics. These substances bore holes through the membranes that surround bacterial cells, causing the cells to disintegrate and die. Bacterial membranes don't mutate often, so they are less likely to become resistant to the animal antibiotics.

Getting Well

► When your doctor prescribes antibiotics, you are usually reminded to finish the entire bottle even if you start to feel better. Call or visit a local pharmacy to investigate why this is so important when you are taking antibiotics.

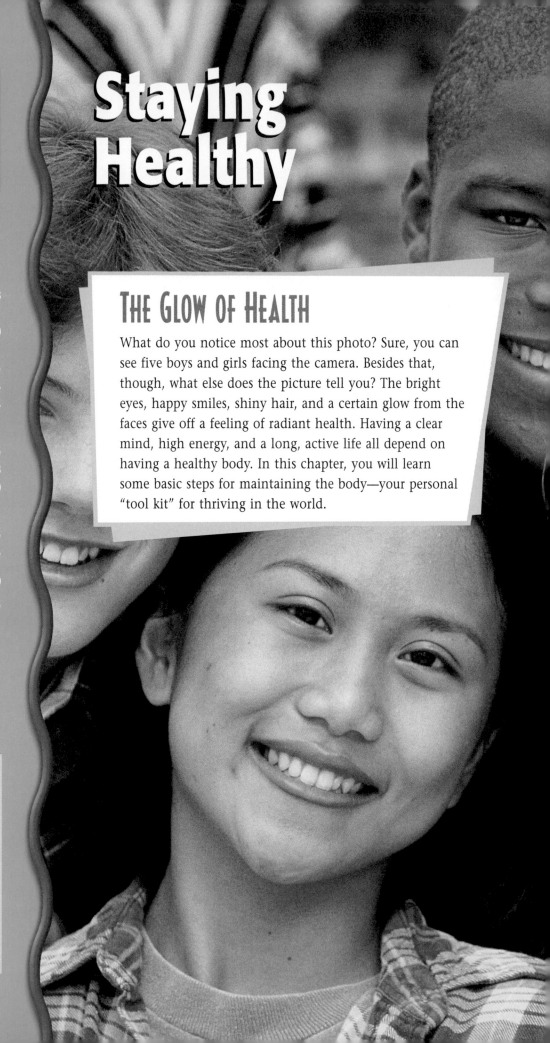

CHAPTER 7

Staying Healthy

Sections

Pre-Reading Questions

1. What types of foods should you eat daily to have a healthy diet? What foods should you avoid?

2. How are drugs helpful? harmful?

3. What everyday habits can help to keep you healthy?

THE GLOW OF HEALTH

What do you notice most about this photo? Sure, you can see five boys and girls facing the camera. Besides that, though, what else does the picture tell you? The bright eyes, happy smiles, shiny hair, and a certain glow from the faces give off a feeling of radiant health. Having a clear mind, high energy, and a long, active life all depend on having a healthy body. In this chapter, you will learn some basic steps for maintaining the body—your personal "tool kit" for thriving in the world.

CONDUCT A SURVEY

How healthy is your class? Collect data and see for yourself.

Procedure

1. Copy the questionnaire below.

2. Circle *yes* or *no* to answer the questions. Do not put your name on the survey.

Analysis

3. Record the data from your survey and the surveys of your classmates in a chart. Count the number of students who answered *yes* to each question. For each question, calculate the percentage of your class that answered *yes*.

4. What things does your class do well? What health habits can be improved?

1. Do you exercise at least three times a week?   Yes   No

2. Do you wear a seat belt every time you ride in a car?   Yes   No

3. Do you eat five or more servings of fruits and vegetables every day?
                        Yes   No

4. Do you use sunscreen to protect your skin when you are outdoors?
                        Yes   No

5. Do you eat a lot of high-fat foods?
                        Yes   No

What We Put into Our Bodies

Terms to Learn

nutrient   mineral
carbohydrate   vitamin
protein   malnutrition
fat

What You'll Do

◆ Identify the six groups of nutrients, and explain their importance to good health.
◆ Use dietary guidelines and the food pyramid to plan a healthy diet.
◆ Understand nutrition information labels.
◆ Explain the dangers of various nutritional disorders.

"You are what you eat." Does this familiar saying mean that you are pizza or candy? Of course not! But, the substances in the pizza and candy enter your body. The protein in the cheese may become part of your hair; the carbohydrates in the crust can give you energy to run your next race. The sugar in the candy can give you a quick energy boost but make you tired later when your blood sugar level drops.

Nutrition

Are you more likely to have potato chips or broccoli for a snack? a candy bar or a banana? If you lean toward foods that are high in sugar and fat, such as potato chips and candy, your food choices probably are not as healthy as they could be. Does that mean you have to cut out all of your favorite foods to eat healthy? No! Broccoli *is* a healthier food than potato chips. But eating only broccoli every day, like the person in **Figure 1,** is not much better than eating only potato chips! Either way, you do not get a balanced diet.

Balancing Act In order to stay healthy, you need to take in more than 40 different substances every day. These substances, or **nutrients,** nourish your body and are essential to life. To get them all, you must eat a wide variety of foods. Nutrients are grouped into six categories: *carbohydrates, proteins, fats, vitamins, minerals,* and *water.* Three of these— carbohydrates, proteins, and fats—provide energy for the body. The energy in these nutrients is measured in units called Calories. The other three nutrients—vitamins, minerals, and water—do not provide energy in Calories but help the body use all of the nutrients properly.

Figure 1 *Eating only one food, even a healthy food, will not give you all the substances your body needs.*

Body Fuel A **carbohydrate** is a chemical composed of one or more simple sugars. Carbohydrates are your body's main source of energy. They help digest fats, lubricate joints, and keep skin, bones, and nails healthy. Plant foods are the major source of carbohydrates.

There are two basic types of carbohydrates: simple and complex. *Simple carbohydrates* are sugars. They are easily digested and give your body quick energy. *Complex carbohydrates* are made of many sugar molecules linked together. They are digested more slowly than simple carbohydrates and give your body long-lasting energy.

Body Builders Protein is found in body fluids, muscle, bone, skin, and all other tissues. **Proteins** are nutrients used to build and repair body parts. Your body makes the proteins it needs, but it must have the necessary building blocks, called *amino acids,* to make them. Your digestive system breaks down the protein in food into individual amino acids that are then used to make new proteins. If your body does not get enough carbohydrates, it can also use proteins for energy.

Some foods, such as poultry, fish, milk, and eggs provide all of the essential amino acids. These food sources are called complete proteins. Incomplete proteins contain only some of the essential amino acids. Most plants are incomplete sources of protein, but eating a variety of plant foods each day will provide all of the amino acids your body needs.

Energy Storage **Fats** are energy-storage nutrients that help the body store some vitamins. Too much fat in the diet has been linked to weight gain, heart disease, and some kinds of cancer. But fats are essential to a balanced diet. They are needed to transport vitamins, produce hormones, keep skin healthy, protect vital organs, and provide insulation. Fats provide more than twice as much energy as proteins and carbohydrates per unit mass.

Simple carbohydrates

Complex carbohydrates

Proteins

Fats

Figure 2 Energy-Producing Nutrients

There are two types of fats, saturated and unsaturated. *Saturated fats* are found in meats, dairy products, coconut oil, and palm oil. Saturated fats are known to raise blood cholesterol levels. *Cholesterol* is a fatlike substance found naturally in the body. Although cholesterol is important to the body, high levels in the blood increase the risk of heart disease. *Unsaturated fats* may help reduce blood cholesterol levels. Your body can make its own saturated fats, but it cannot make certain unsaturated fats. You must get these from your diet. Vegetable oils and fish contain unsaturated fats.

Flushing the System A human cannot survive for more than a few days without water. Your body is about 70 percent water. Water is in every cell and every kind of tissue. Water's three main functions are to transport substances, regulate temperature, and provide lubrication. You should drink 8 to 10 glasses of water daily, as shown in **Figure 3.** You also get water from the other liquids you drink and the foods you eat. Fresh fruits and vegetables, juices, soups, and milk contain large amounts of water.

Small Necessities **Minerals** are elements that are essential for good health. Six minerals are needed in large amounts: calcium, chloride, magnesium, phosphorus, potassium, and sodium. There are at least 12 minerals that are required in very small amounts. These include fluorine, iodine, iron, and zinc. If you eat a balanced diet, you should get all of the minerals you need. Calcium and magnesium are necessary for strong bones and teeth. Magnesium and sodium help the body use proteins. Potassium is needed to regulate your heartbeat and produce muscle movement, and iron is necessary for red blood cell production.

Figure 3 *You need to drink about eight glasses of water a day. When you exercise, you need even more.*

Body Controllers **Vitamins** are organic compounds that control many body functions. Most vitamins cannot be made by the body, so you have to get them from food. The following table provides information about the 13 essential vitamins.

The Essential Vitamins

Vitamin	What it does	Where you get it
A	keeps skin and eyes healthy; builds strong bones and teeth	yellow and orange fruits and vegetables; dark, leafy greens; meat; and milk
B_1 (thiamine)	helps body use carbohydrates; helps nerves and heart function	meats, whole grains, beans, peas, nuts, and seafood
B_2 (riboflavin)	helps cells use carbohydrates and oxygen; keeps skin and eyes healthy	dairy products, fruits, whole grains, eggs, leafy vegetables, and poultry
B_3 (niacin)	helps body use carbohydrates; helps cells use oxygen; helps digestion	meats, peanuts, whole grains, peas, and beans
B_6	helps body use proteins, carbohydrates, and fats	poultry, fish, meat, eggs, potatoes, avocados, and bananas
B_{12}	keeps blood and nerves healthy	meats, poultry, eggs, fish, and milk
Folic acid (a B vitamin)	helps red blood cell formation	leafy greens, peas, beans, nuts, whole grains, liver, and oranges
Pantothenic acid (a B vitamin)	helps body use protein, carbohydrates, and fat; keeps body tissues healthy	meats, fish, whole grains, beans, peas, eggs, and corn
Biotin (a B vitamin)	helps body use protein, carbohydrates, some B vitamins, and fat	eggs, milk, meats, nuts, peas, beans, and whole grains
C	strengthens blood vessels and connective tissue; helps the body absorb iron; helps the body fight disease	citrus fruits; dark, leafy greens; broccoli; peppers; cabbage; tomatoes; potatoes; and strawberries
D	builds strong bones and teeth; helps the body use calcium and phosphorus	sunlight, enriched milk, eggs, and fish
E	protects red blood cells from destruction; needed for some enzymes to work	oils, fats, eggs, whole grains, wheat germ, liver, and leafy greens
K	assists with blood clotting	leafy greens, tomatoes, and potatoes

SECTION REVIEW

1. Name the six groups of nutrients, and explain why each is important to the body.

2. If vitamins and minerals do not supply energy, why are they important to a healthy diet?

3. **Applying Concepts** Name some of the nutrients that can be found in a glass of milk.

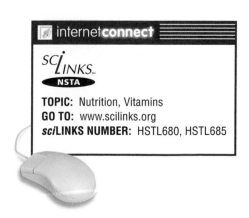

internet**connect**

sci**LINKS**
NSTA

TOPIC: Nutrition, Vitamins
GO TO: www.scilinks.org
*sci*LINKS NUMBER: HSTL680, HSTL685

Eating for Good Health

Now you know what nutrients you need for good health. But how can you be sure to get them all in the right amounts? To begin, keep in mind that most teenage girls need about 2,200 Calories a day and most boys need 2,800 Calories. Since different foods contain different nutrients, *where* you get your Calories is as important as *how many* you get. The food pyramid below can help you make good food choices.

The Food Pyramid

The U.S. Department of Agriculture and the Department of Health and Human Services developed the food pyramid to help Americans make healthy food choices. The food pyramid divides foods into six groups. It shows how many daily servings you need from each group and gives examples of foods for each. This food pyramid also provides sample serving sizes for each group. Within each group, the food choices are up to you. You can eat anything you want. By following the food pyramid, you can achieve a healthy, balanced diet.

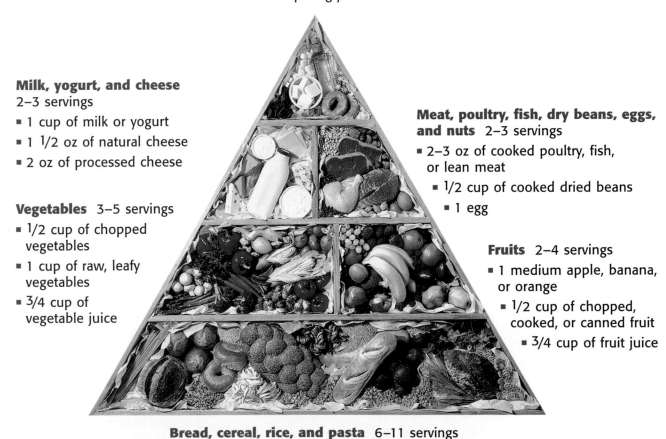

Fats, oils, and sweets
Use sparingly.

Milk, yogurt, and cheese
2–3 servings
- 1 cup of milk or yogurt
- 1 1/2 oz of natural cheese
- 2 oz of processed cheese

Vegetables 3–5 servings
- 1/2 cup of chopped vegetables
- 1 cup of raw, leafy vegetables
- 3/4 cup of vegetable juice

Meat, poultry, fish, dry beans, eggs, and nuts 2–3 servings
- 2–3 oz of cooked poultry, fish, or lean meat
  - 1/2 cup of cooked dried beans
  - 1 egg

Fruits 2–4 servings
- 1 medium apple, banana, or orange
  - 1/2 cup of chopped, cooked, or canned fruit
  - 3/4 cup of fruit juice

Bread, cereal, rice, and pasta 6–11 servings
- 1 slice of bread
- 1 oz of ready-to-eat cereal
- 1/2 cup of rice or pasta
- 1/2 cup of cooked cereal

How to Read a Food Label

Packaged foods are required by law to have nutrition information labels. The illustration below shows a nutrition information label for a box of macaroni and cheese. Reading labels will help you make healthy eating choices.

Some of the nutrients found in each serving are listed on the label. The daily values shown are based on a 2,000-Calorie-per-day diet. You will need to calculate your own daily values based on your personal Calorie needs.

The **nutrition facts panel** ▶ contains the serving size, the number of servings per container, and the number of Calories per serving.

The **daily value** represents ▶ how much of each nutrient you need each day.

Nutrition Facts

Serving Size 1 cup (59g)
Serving Per Container 2

Amount Per Serving	Prepared
Calories	290
Calories from Fat	90
	% Daily Value
Total Fat 10g	14%
Saturated Fat 3.5g	16%
Cholesterol 10mg	39%
Sodium 30mg	39%
Total Carbohydrate 41g	14%
Dietary Fiber less than 1g	3%
Sugars 4g	
Protein 10g	12%
Vitamin A	8%
Vitamin C	0%
Calcium	15%
Iron	8%
Thiamine	30%
Riboflavin	15%
Niacin	15%
Folic Acid	25%

The second part of a food label is the **list of ingredients.** It begins with the ingredient used in the greatest amount and ends with the ingredient used in the least amount.

Ingredients: Enriched Macaroni Product (Wheat Flour, Niacin, Ferrous Sulfate (Iron), Thiamine Mononitrate (Vitamin B1), Riboflavin (Vitamin B2), and Folic Acid), Milk, Cheddar Cheese (Pasteurized Milk, Cheese Culture, Salt, Enzymes), Whey, Margarine (Partially Hydogenated Soybean Oil, Water, Soy Lecithin, Mono- and Diglycerides, Beta Carotene for Color, Vitamin A Palmitate), Salt, Cornstarch, Yeast Extract, Lactic Acid, Sodium Citrate, Spices, Annatto (for color).

MATH BREAK

What Percentage?

Use the nutrition information label on this page to answer the following questions for yourself, based on the Calorie needs of teenagers described on the previous page.

1. What percentage of your daily Calorie needs does one serving of macaroni and cheese provide?

2. The recommended daily value of fat is 72 g for teenage girls and 90 g for teenage boys. What percentage of the daily recommended fat value is provided by one serving of packaged macaroni and cheese?

✓ **Self-Check**

For breakfast, you eat 1 cup of hot cereal with a banana and 1 cup of milk. What servings from the food pyramid have you eaten? *(See page 212 to check your answer.)*

A Healthy Diet

1. Use the recommended daily Calories for teenagers and the **table** below to estimate how many servings from each group in the food pyramid you should eat daily.

2. Create a menu for 2 days that includes the correct number of servings from each group in the food pyramid.

3. Compare this diet with your normal diet. How different are they? What can you do to improve your diet?

Daily Number of Servings

Food group	2,200 Calories	2,800 Calories
Bread	9	11
Fruit	3	4
Vegetable	4	5
Dairy	3	3
Meats	6 oz	7 oz

TRY at HOME

internet**connect**

SC*i*LINKS
NSTA

TOPIC: Food Pyramids, Nutritional Disorders
GO TO: www.scilinks.org
*sci*LINKS NUMBER: HSTL690, HSTL695

Nutritional Disorders

Unhealthy eating habits can cause nutritional disorders. For example, malnutrition can result from consuming too few Calories or too few of the right nutrients. Eating too many Calories or too many of the wrong nutrients can also cause malnutrition. **Malnutrition** occurs when you do not consume the right combination of nutrients.

Anorexia Nervosa and Bulimia   Anorexia nervosa and bulimia can lead to malnutrition. Many of the people who suffer from anorexia nervosa and bulimia are teenage girls. *Anorexia nervosa* is an eating disorder characterized by self-starvation and an intense fear of gaining weight. This can cause weak bones, low blood pressure, and heart problems. *Bulimia* is a disorder characterized by binge eating followed by induced vomiting. Sometimes people suffering from bulimia also use laxatives and diuretics to rid the body of food and water. Bulimia can damage the teeth and digestive system and can also lead to kidney or heart failure. These disorders can both be fatal, but they can be cured with medical help.

Obesity   *Obesity* is a condition characterized by an extremely high percentage of body fat. Eating too many foods from the top of the food pyramid and having an inactive lifestyle that involves little exercise can contribute to obesity. Obesity increases the risk of high blood pressure, heart disease, and diabetes.

SECTION REVIEW

1. What information is found on a nutrition information label?

2. How do anorexia nervosa and bulimia differ?

3. **Applying Concepts**   How can someone who is obese suffer from malnutrition?

SECTION 2
READING WARM-UP

Terms to Learn

drug   alcoholism
addiction   narcotics
nicotine

What You'll Do

◆ Distinguish between the positive and negative uses of drugs.
◆ Explain the hazards of tobacco, alcohol, and illegal drugs.

Risks of Alcohol and Other Drugs

You see them in movies and on television. You read about them in magazines. You hear about them in music and in your school. You are exposed to information, and misinformation, about tobacco, alcohol, and other drugs almost every day.

What Is a Drug?

A **drug** is any chemical substance that causes a physical or emotional change in a person. Drugs come in many forms, as shown in **Figure 4.** They can be pills, powders, fumes, liquids, or creams. Some drugs enter the body through the skin, and others are swallowed, inhaled, or injected.

When used safely and correctly, legal drugs can help your body heal a variety of ailments from athlete's foot to pneumonia. They can also provide relief from pain, congestion, and other symptoms. When used illegally or improperly, however, drugs can become killers.

What Effects Do Drugs Have?

Different drugs have different effects. One way to classify drugs is by function. *Analgesics,* like aspirin, are pain-relieving drugs. Drugs that relax muscles and help people sleep are *sedatives.* *Antibiotics* fight bacterial infections, and *antihistamines* help control symptoms of allergies, asthma, and colds.

Another way to classify drugs is by their effect on the central nervous system. *Stimulants* speed up the action of the central nervous system and may cause a person to feel more alert. *Depressants* have the opposite effect. They slow down body functions and may reduce a person's alertness.

Figure 4 *All of these products contain drugs.*

Self-Check

How are tolerance, addiction, and withdrawal symptoms related? *(See page 212 to check your answer.)*

Dependence and Addiction Regular use of some drugs can cause the body to develop *tolerance*. This means larger and larger doses of the drug are needed to get the same effect. **Addiction** is physical dependence on a drug. When a person is addicted to a drug, the body has a chemical need for the drug. If the body doesn't receive the drug, withdrawal symptoms may occur. These include nausea, vomiting, pain, and other physical symptoms. Once addicted, it is very difficult to stop taking a drug.

Sometimes dependence on a drug is not due to a physical need. Some people form *psychological dependence* on a drug and feel powerful cravings for it.

Types of Drugs

There are many kinds of drugs and many ways to use them. Some drugs are obtained from plants, and some are made synthetically. You can buy some drugs off the shelf, while others must be taken under the supervision of a doctor. Some drugs are illegal to buy, sell, or even possess.

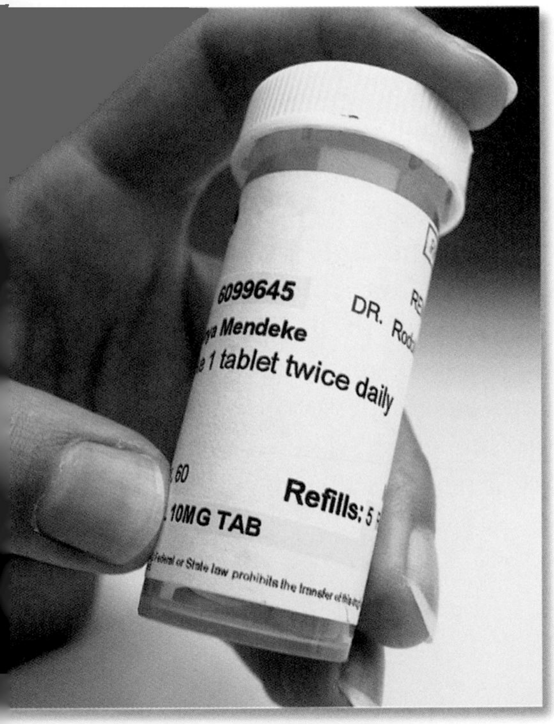

Over-the-Counter and Prescription Drugs An over-the-counter drug can be purchased legally without a doctor's prescription. A prescription is a note written by a doctor to allow a patient to buy a medicine. It specifies the drug, directions for use, and the amount of the drug to be used.

Many over-the-counter and prescription drugs are powerful healing agents. However, some of these drugs also produce unwanted side effects. Side effects are uncomfortable symptoms such as nausea, headaches, drowsiness, or more serious problems caused by a drug.

Whether purchased with or without a prescription, all drugs must be used with care. Each year about 75,000 people in the United States become ill or die from the misuse of drugs. Information on proper use can be found on the label. **Figure 5** shows an example of a prescription drug label. The table below gives some drug safety tips.

Figure 5 *Prescription drug labels provide instructions for use and list possible side effects.*

Drug Safety Tips

- Never take another person's prescription medicine.
- Read the label before each use. Always follow the instructions on the label and those provided by your doctor or pharmacist.
- Do not take more or less medication than prescribed.
- Consult a doctor if you have any side effects.
- Throw away leftover and out-of-date medicines.

Herbal Medicines Information about medicinal herbs has been handed down by word-of-mouth for centuries. Some plants contain chemicals with important healing properties. However, these herbs are drugs and should be used carefully. **Figure 6** shows some medicinal herbs.

Tobacco About 50 million people in the United States—one-third of all adults—smoke. Smoking has serious health risks, and the nicotine in cigarettes is addictive. **Nicotine** is a chemical stimulant in tobacco that increases heart rate and blood pressure. Many smokers also experience a loss of appetite and a decrease in physical endurance. Smoking increases the chances of lung cancer by 10 times, and it has also been linked to other cancers, emphysema, chronic bronchitis, and heart disease. About 400,000 people die from smoking-related illnesses each year. **Figure 7** shows one of the effects of smoking.

Smokeless tobacco is also a health hazard. Nicotine is absorbed through the lining of the mouth, and the amount of nicotine that reaches the blood can be the same as for a smoker. Smokeless tobacco increases the risk of several types of cancer, including mouth and throat cancer. It also causes gum disease and discoloration of the teeth.

Figure 6 *Some herbs can be purchased in health food stores. Medicinal herbs should always be used with care.*

Figure 7 *Cilia in your airways cleanse debris from the air you breathe and prevent debris from entering your lungs. Compare the cilia from the lungs of a nonsmoker, on the left, with those of a smoker, on the right.*

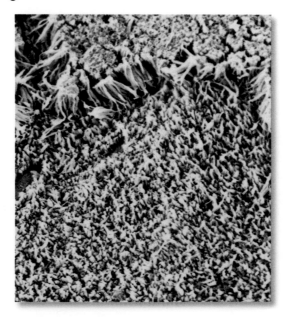

Figure 8 *On average, there is one alcohol-related fatality every 31 minutes.*

Alcohol Alcohol depresses the central nervous system and causes relaxation and memory loss. Excessive use of alcohol can damage the liver, pancreas, brain, nerves, and cardiovascular system. In very large quantities, alcohol can cause respiratory failure and even death. In addition, alcohol is a factor in more than half of all accidental deaths, suicides, and murders. **Figure 8** shows one example of an alcohol-related accident. It is illegal in the United States for people under the age of 21 to use alcohol.

About 10 million alcohol users are alcoholics. They suffer from **alcoholism,** which means that they are physically and psychologically dependent on alcohol. Alcoholism is considered a disease, and genetic factors are thought to influence a person's tendency to become an alcoholic.

Marijuana Marijuana is an illegal drug made from the Indian hemp plant. Marijuana produces a mind-altering effect that varies from user to user. It may cause a relaxed feeling or may increase anxiety. Marijuana slows reaction time, impairs thinking, and causes a loss of coordination.

Cocaine Cocaine and its more purified form, crack, are stimulants made from the South American coca plant. Both drugs are illegal and highly addictive, and users can become dependent on them in a very short time. Cocaine and crack produce feelings of intense excitement followed by anxiety and depression. Both drugs increase heart rate and blood pressure and can cause heart attacks even among first-time users.

Hallucinogens Hallucinogens (huh LOO si nuh juhnz) distort the senses and cause changes in mood and thought processes. Users have hallucinations (huh LOO si NAY shuhnz), which means they see and hear things that are not real. Some hallucinations are extremely frightening and can cause people to respond violently. LSD and PCP, or "angel dust," are two powerful and illegal hallucinogens. Sniffing some glues and solvents also causes hallucinations and can cause serious brain damage.

MATH BREAK

Deadly Averages

Approximately 2,200 people between the ages of 16 and 20 die in alcohol-related crashes each year. On average, how many die every day?

Narcotics **Narcotics** are drugs made from opium. Some narcotics are used to treat severe pain and are legal when prescribed by a doctor. But many narcotics are illegal.

One illegal narcotic is heroin. Heroin is one of the most addictive drugs known, and large doses can lead to death. Because it is so strongly addictive, users must continue taking the drug in order to prevent painful withdrawal symptoms. Heroin is usually injected, and users often share needles that are contaminated. Therefore, heroin users have a high risk of infecting themselves with diseases like hepatitis and AIDS.

Designer Drugs There are many other illegal drugs. Examples include inhalants, barbiturates (downers), amphetamines (uppers), and "designer" drugs, which are produced by making small changes to existing drugs. Some inhalants, barbiturates, and amphetamines are legal if prescribed by a doctor.

✓ **Self-Check**

1. What is a hallucination?
2. List three dangers of heroin use.

(See page 212 to check your answers.)

Drug Use and Drug Abuse

A drug user takes a drug to prevent or improve some medical condition. The drug user obtains the drug legally and uses the drug properly. A drug abuser does *not* take a drug to relieve a medical condition. The drug abuser may take drugs for the temporary good feelings they produce, to escape from problems, or to belong to a group. The drug is often obtained illegally, and it is often taken without knowledge of the strength or purity of the drug.

Drug Abuse: How Does It Start? Nicotine, alcohol, and marijuana are called *gateway drugs* because they are usually the first drugs a person tries. The abuse of other more dangerous drugs may follow the abuse of gateway drugs, as illustrated in **Figure 9.** Most teenagers who start smoking cigarettes, drinking alcohol, or using marijuana do not realize that these drugs are addictive or that they can seriously harm their health.

Peer pressure is the reason most young people begin to use drugs. Teenagers may feel the need to drink, smoke, or try marijuana in order to make friends or to avoid being ridiculed or threatened. Many young people feel that "everyone" is using drugs. Because drug abusers often stand out, the fact that many teenagers do not abuse drugs is sometimes hard to see.

Figure 9 *Nicotine, alcohol, and marijuana are called gateway drugs.*

Many people who start using drugs do not recognize the dangers. Misinformation about drugs is everywhere. Several common drug myths are discussed below.

Drug Myths

Myth	Reality
"It's only alcohol, not drugs."	Alcohol is a mood-altering and mind-altering drug. It affects the central nervous system and is addictive.
"I won't get hooked on one or two cigarettes a day."	Addiction is not related to the amount of a drug used. Some people become addicted after using a drug once or twice.
"I can quit any time I want."	Addicts may quit and return to drug usage many times. Their inability to stay drug free shows how powerful the addiction is.

Activity

Write yourself a letter. Tell yourself why you should stay drug free. You may wish to refer to people, goals, activities, or values that are important to you. Put your letter in an envelope, and keep it in a safe place. If you ever find yourself thinking about using drugs, take out your letter and read it.

TRY at HOME

Signs of Use People who begin using drugs generally undergo emotional, physical, and behavioral changes. Teenagers may have problems with school or family. Changes in personality or physical appearance may occur. However, remember that many young people do not take drugs. These changes could be signs of other problems or normal age-related changes.

Getting Off Drugs The first step to quitting drugs is to admit to being a drug abuser and to decide to stop. When quitting drugs, it is important for the addicted person to get proper medical and psychological treatment. Getting off drugs can be extremely difficult. Withdrawal symptoms may be painful, and even after the symptoms are gone, powerful cravings for a drug continue. This is even true of smoking.

SECTION REVIEW

1. What is the difference between a prescription drug and an over-the-counter drug?

2. How does addiction occur?

3. **Analyzing Relationships** How are nicotine, alcohol, heroin, and cocaine similar? How are they different?

SECTION 3

READING WARM-UP

Terms to Learn

hygiene   stress
aerobic exercise

What You'll Do

◆ Describe four important aspects of good hygiene.
◆ Explain why exercise and sleep are important to good health.
◆ Describe methods of handling stress.
◆ List ways to stay safe at home, on the road, and outdoors.

Healthy Habits

Do you like playing sports? acting in plays? going to the movies? swimming in the ocean? No matter what you enjoy, the better your health, the easier it will be to take part in the things you like to do. Keeping yourself healthy is a daily task. If you have healthy habits, you are likely to stay healthy for a long time.

Hygiene and Posture

Hygiene refers to methods of preserving and protecting your health. It sounds simple, but washing your hands is the best way to prevent the spread of disease and infection. You should always wash your hands after using the bathroom and before and after eating or preparing food.

Taking care of your skin, hair, and teeth are other important aspects of good hygiene. Using sunscreens can help prevent sunburn, wrinkles, and skin cancer. Shampoo your hair regularly. To prevent cavities and keep your teeth and gums healthy, eat a healthy diet, brush at least twice a day, and floss at least once daily. Get regular dental exams, and replace your toothbrush about every 3 months.

Posture   Posture is also important to health. Good posture helps you look and feel good. Bad posture strains your muscles and ligaments and makes breathing difficult. To have good posture when you stand, imagine a vertical line passing through your ear, shoulder, hip, knee, and ankle, as shown in **Figure 10.** When working at a desk, pull your chair forward and plant your feet firmly on the floor.

Figure 10 *A slumped posture strains your lower back.*

Staying Healthy   **161**

Exercise and Rest

Imagine that school starts at 7:30 A.M. You drag yourself out of bed at 6:00 every morning. You are tired, but somehow you make it through the school day. In the afternoon, you attend play rehearsal. When you get home, you take a break, eat dinner, and start your homework at 8:00 P.M. When you finish, you do not feel ready to sleep, so you watch TV, then read in bed, and finally, at 10:30, you drift off to sleep.

Keeping your body healthy requires giving it plenty of exercise and plenty of rest. If you have a schedule like the one above, you do not get enough of either.

Figure 11 *To stay with aerobic exercise, it is important to choose an activity you enjoy.*

Exercise Regular aerobic exercise at least three times a week is critical to good health. **Aerobic exercise** is vigorous, sustained exercise of the whole body for 20 minutes or more. Walking, running, swimming, and biking are all examples of aerobic exercise. For another example, see **Figure 11.**

Aerobic exercise increases the heart rate. As a result, more oxygen is taken in and distributed throughout the body. Over time, aerobic exercise strengthens the heart, lungs, and bones. It burns Calories, helps your body conserve nutrients, and aids digestion. It also gives you more energy and stamina. In other words, aerobic exercise protects your physical and mental health, and it's free! What a deal!

BRAIN FOOD

Only half of a dolphin's brain sleeps at a time. A full, deep sleep would be a problem because dolphins must come to the surface for air every 20–30 seconds.

Sleep Believe it or not, teenagers actually need more sleep than younger children. Do you ever fall asleep in class, like the girl in **Figure 12,** or feel tired in the middle of the afternoon? If so, you may not be getting enough sleep. Scientists say that teenagers need about 9.5 hours of sleep each night.

At night, the body goes through several cycles of progressively deeper sleep, with periods of lighter sleep in between. If you do not sleep long enough, you will not enter the deepest, most restful period of sleep.

Figure 12 *If you fall asleep easily during the day, you are probably not getting enough sleep.*

Coping with Stress

You have a big soccer game tomorrow. Are you excited and ready for action? You got a low grade on your math test. Are you upset or angry? The game and the test are causing you stress. **Stress** is the physical and mental response to pressure.

Some stress is a normal part of life, as shown in **Figure 13.** Stress stimulates your body to prepare for difficult or dangerous situations. However, sometimes you may have no outlet for the stress, and it builds up. Excess stress is harmful to your health and can decrease your ability to carry out your daily activities.

You may not even realize you are stressed until your body reacts. Perhaps you get a headache, have an upset stomach, or lie awake at night. You might feel tired all the time or begin an old nervous habit, such as nail-biting. You may become irritable or resentful. All of these things can be signs of too much stress.

Figure 13 *Working under stress can increase an athlete's ability to perform well.*

What to Do Different people are stressed by different things. Once you identify the source of the stress, you can find ways to deal with it. Here are some ideas for handling stress.

- Make a list of all the things you would like to get done, and rank them in order of importance. Do the most important things first.

- Exercise regularly, and get enough sleep.

- Pet a friendly animal.

- If you cannot remove a stressor, spend some quiet time alone or practice deep breathing or other relaxation techniques.

- Share your problems. Talk things over with someone you trust.

Stress SOS

Your mother is out of town helping your grandfather move into a nursing home. Your little brother has been snooping in your room. You have a history project due in two days, an oral report due in English, and quizzes in both math and science! You are ready to scream! What are some healthy ways to handle all the stress you are feeling?

Injury Prevention

Have you ever fallen off your bike or sprained your ankle? Maybe you avoided injury, or maybe you ended up in the emergency room. Accidents happen, and they can cause injury and even death. It is impossible to prevent all accidents, but you can decrease your risk by using your common sense and following basic safety rules.

Safety at Home Many accidents can be avoided. **Figure 14** below shows tips for safety around the house.

Figure 14 Home Safety Tips

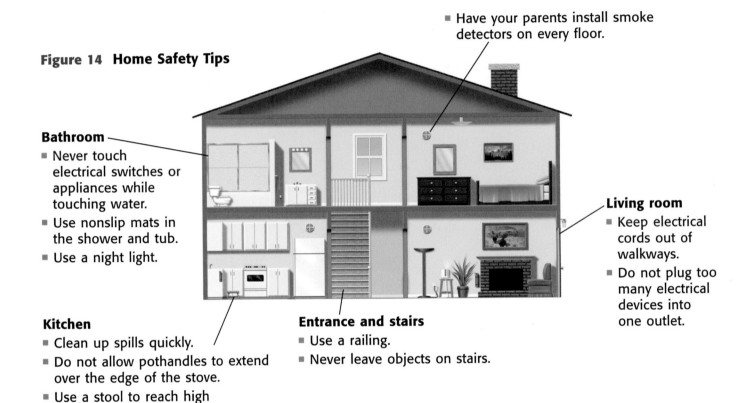

- Have your parents install smoke detectors on every floor.

Bathroom
- Never touch electrical switches or appliances while touching water.
- Use nonslip mats in the shower and tub.
- Use a night light.

Living room
- Keep electrical cords out of walkways.
- Do not plug too many electrical devices into one outlet.

Kitchen
- Clean up spills quickly.
- Do not allow pothandles to extend over the edge of the stove.
- Use a stool to reach high shelves.
- Keep grease and drippings away from open flames.

Entrance and stairs
- Use a railing.
- Never leave objects on stairs.

Safety Outdoors Always dress appropriately for the weather and the activity. Never hike or camp alone. Tell someone where you are going and when you expect to return. If you do not bring water from home, be sure to purify any water you drink in the wilderness.

Learn how to swim. It could save your life! Never swim alone, and do not dive into shallow water or water of unknown depth. When in a boat, wear a life jacket. If a storm threatens, get out of the water and seek shelter.

Safety on the Road In the car, always wear a seat belt, even if you are traveling only a short distance. Most car accidents happen close to home, and seat belts save lives. When riding a bicycle, always wear a helmet. Ride with traffic, and obey all traffic rules. Be sure to signal when stopping or turning.

When Accidents Happen

No matter how well we practice safety measures, accidents can still happen. What should you do if a friend chokes on food and cannot breathe? What if he is stung by a bee and has a violent allergic reaction?

Call for Help Once you've checked for other dangers, call for medical help immediately, like the person in **Figure 15**. In most communities you can dial 911. Speak slowly and clearly. Give the complete address and a description of the location. Describe the accident, the number of people injured, and the types of injuries. Ask what to do and listen carefully to the instructions. Let the other person hang up first to be sure they have no more questions or instructions.

Figure 15 *When calling 911, stay calm and listen carefully to what the dispatcher tells you.*

Learn First Aid You may want to learn more about what to do in an emergency by taking a first-aid course or a CPR course, as shown in **Figure 16**. *CPR* is a lifesaving procedure designed to revive a person who is not breathing and has no heartbeat. If you are over 12 years old, you can become certified in both CPR and first aid. Some babysitting courses also provide basics in first aid and are a good idea for anyone who cares for young children. The American Red Cross, community organizations, and local hospitals offer these classes. You should not attempt any lifesaving procedure unless you have been trained.

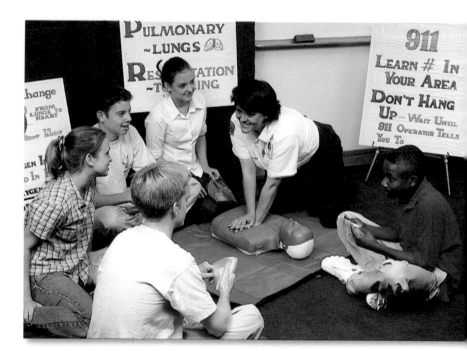

Figure 16 *These teenagers are taking a CPR course to prepare themselves for emergency situations.*

SECTION REVIEW

1. What is aerobic exercise? Give three examples.

2. What should you do when calling for help in a medical emergency?

3. What is hygiene, and how does it help you stay healthy?

4. **Applying Concepts** What situations are causing excess stress in your life right now? What can you do to help relieve the stress you are feeling?

Discovery Lab

To Diet or Not to Diet

There are six main classes of foods that we need in order to keep our bodies functioning properly: water, vitamins, minerals, carbohydrates, fats, and proteins. In this activity, you will investigate the importance of a well-balanced diet in maintaining a healthy body. Then you will create a poster or picture that illustrates the importance of one of the three energy-producing nutrients—carbohydrates, fats, and proteins.

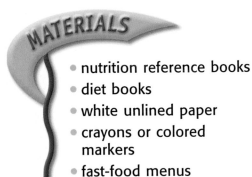

MATERIALS

- nutrition reference books
- diet books
- white unlined paper
- crayons or colored markers
- fast-food menus (optional)

Procedure

1 In your ScienceLog, create a table like the one below. Research in the library, on nutrition labels, in nutrition or diet books, or on the Internet to find the information you need to fill out the chart.

2 Choose one of the foods you have learned about in your research, and create a poster or picture that describes its importance in a well-balanced diet.

Analysis

3 Based on what you have learned in this lab, how might you change your eating habits to have a well-balanced diet? Does the nutritional value of foods concern you? Why or why not? Write your answers in your ScienceLog, and explain your reasoning.

Nutrition Data Table			
	Fats	**Carbohydrates**	**Proteins**
Found in which foods			
Functions in the body			
Consequences of deficiency			

DO NOT WRITE IN BOOK

Discovery Lab

Keep It Clean

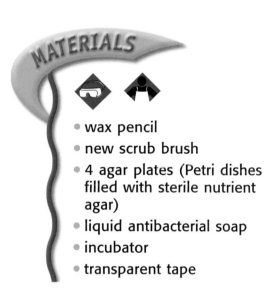

One of the best ways of preventing the spread of bacterial and viral infections is to wash your hands with soap and water frequently. Many companies advertise that their soap ingredients will destroy bacteria normally found on the body. In this activity, you will investigate how effective antibacterial soaps are at killing bacteria.

MATERIALS

- wax pencil
- new scrub brush
- 4 agar plates (Petri dishes filled with sterile nutrient agar)
- liquid antibacterial soap
- incubator
- transparent tape

Procedure

1. Label four agar plates with "Control," "Unwashed," "No soap," and "Soap." Open the "Control" plate for 1 minute. Cover the dish, and leave it closed for the rest of the experiment.

2. Carefully press different surfaces of your unwashed hands on the "Unwashed" plate. Replace the cover immediately.

3. Hold your right hand under running water for two minutes. During this time, ask your partner to scrub all surfaces of your hand with the scrub brush. When you are finished, your partner should turn off the water and open the plate marked "No soap." Carefully press the same surfaces of your right hand on the plate, as you did in step 2.

4. Repeat step 3, using liquid antibacterial soap on your left hand. Do not touch anything before pressing your left hand on the plate marked "Soap."

5. Tape the lid of each plate to its bottom. Incubate the plates upside down overnight at 37°C.

6. Remove the plates from the incubator and turn them right side up. Check each plate for the presence of bacterial colonies, and count the number of colonies present on each plate.
Caution: Do not remove any of the lids.

Analysis

7. Which plate contained the most growth? Which contained the least?

8. Does water alone effectively kill bacteria? Explain.

Chapter Highlights

SECTION 1

Vocabulary
nutrient *(p. 148)*
carbohydrate *(p. 149)*
protein *(p. 149)*
fat *(p. 149)*
mineral *(p. 150)*
vitamin *(p. 151)*
malnutrition *(p. 154)*

Section Notes

• Carbohydrates, proteins, fats, vitamins, minerals, and water are the six types of nutrients that are essential to life.

• Teenage girls need about 2,200 Calories a day. Boys need about 2,800. It is important to obtain Calories from a variety of foods.

• The food pyramid gives information for eating a balanced diet.

• A nutrition information label on a packaged food lists serving size, Calorie and nutrient content, and the ingredients.

• Anorexia nervosa, bulimia, and obesity can lead to malnutrition.

SECTION 2

Vocabulary
drug *(p. 155)*
addiction *(p. 156)*
nicotine *(p. 157)*
alcoholism *(p. 158)*
narcotics *(p. 159)*

Section Notes

• A drug is any chemical substance that causes a physical or emotional change in a person.

• Some drugs can cause addiction or psychological dependence.

• Over-the-counter and prescription drugs are legal drugs, many of which are powerful healing agents. Some also have side effects.

• Alcohol and nicotine are legal drugs for adults. Both are addictive and hazardous to your health.

☑ Skills Check

Math Concepts

PERCENTAGE If your recommended daily value of fat is 72 g and you eat a candy bar that has 12 g of fat, what percentage of your daily value of fat have you eaten? To find the percentage, divide the grams of fat in the candy bar by the daily value. Then multiply by 100.

$$12 \text{ g} \div 72 \text{ g} = 0.17 \times 100\% = 17\%$$

The fat in the candy bar is 17 percent of your total daily value.

Visual Understanding

NUTRITION INFORMATION LABELS All packaged foods in the United States are required to have a nutrition label like the one on page 153. On this label you can see that the serving size is 1 cup and that there are two servings in the container. The number of Calories in a serving and the number of Calories from fat are listed. You will also see the percent daily value for a number of nutrients that can be found in one serving. These percentages are based on 2,000 Calories a day.

SECTION 2

- A drug abuser takes a drug, often illegally, for non-medical reasons.

- Drug abuse oftens begins with the use of tobacco, alcohol, or marijuana—the gateway drugs.

- Getting off drugs requires admitting addiction, deciding to stop, going through withdrawal symptoms, and experiencing cravings for the drug.

SECTION 3

Vocabulary

hygiene *(p. 161)*
aerobic exercise *(p. 162)*
stress *(p. 163)*

Section Notes

- Good hygiene is essential to good health.

- Having a healthy body requires getting regular aerobic exercise and getting enough sleep. Posture is also important to good health.

- Some stress is a normal and necessary part of life. Too much stress can result in poor health.

- It is possible to prevent some accidents by using common sense and following basic rules of safety.

- In an emergency, call for help as soon as possible. Do not attempt any lifesaving procedure for which you are not trained.

---

internetconnect

GO TO: go.hrw.com

Visit the **HRW** Web site for a variety of learning tools related to this chapter. Just type in the keyword:

KEYWORD: HSTBD7

GO TO: www.scilinks.org

Visit the **National Science Teachers Association** on-line Web site for Internet resources related to this chapter. Just type in the *sci*LINKS number for more information about the topic:

TOPIC: Nutrition	sciLINKS NUMBER: HSTL680
TOPIC: Vitamins	sciLINKS NUMBER: HSTL685
TOPIC: Food Pyramids	sciLINKS NUMBER: HSTL690
TOPIC: Nutritional Disorders	sciLINKS NUMBER: HSTL695
TOPIC: Drug and Alcohol Abuse	sciLINKS NUMBER: HSTL700

Chapter Review

USING VOCABULARY

To complete the following sentences, choose the correct term from each pair of terms listed below:

1. __?__ are linked to high blood cholesterol levels. *(Saturated fats or Unsaturated fats)*

2. Physical dependence on a drug is called __?__. *(addiction or tolerance)*

3. The __?__ divides foods into six groups and gives a recommended number of servings for each. *(nutrition information label or food pyramid)*

4. One of the characteristics of __?__ is binge eating. *(bulimia or anorexia nervosa)*

5. A person who uses drugs for their intended purpose and with the proper dosage is a __?__. *(drug abuser or drug user)*

UNDERSTANDING CONCEPTS

Multiple Choice

6. Which of the following is *not* a function of water in the body?
    a. transport substances
    b. regulate temperature
    c. provide Calories
    d. provide lubrication

7. Side effects of over-the-counter and prescription medicines may include
    a. nausea.
    b. headaches.
    c. drowsiness.
    d. All of the above

8. Which of the following nutrients does *not* provide energy for the body?
    a. carbohydrates
    b. vitamins
    c. fats
    d. proteins

9. What are the effects of nicotine on the body?
    a. liver damage and decrease in physical endurance
    b. loss of appetite and decrease in physical endurance
    c. loss of appetite and brain damage
    d. liver damage and loss of appetite

10. Which of the following statements about drugs is true?
    a. All drugs are illegal.
    b. Smoking just one or two cigarettes is safe for anyone.
    c. Alcohol is not a drug.
    d. Withdrawal symptoms may be painful.

11. Aerobic exercise does *not*
    a. burn calories.
    b. increase the heart rate.
    c. strengthen the heart, lungs, and bones.
    d. make you weak.

12. When talking to a 911 operator, you should *not*
    a. describe the accident.
    b. ask what to do.
    c. hang up before the dispatcher hangs up.
    d. speak slowly and clearly.

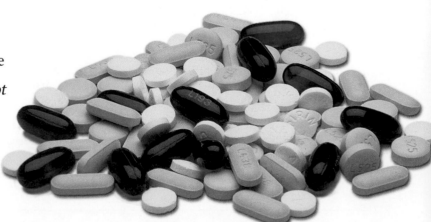

Short Answer

Answer each of the following questions with a few sentences:

13. Are all narcotics illegal? Explain.

14. What are the three types of nutrients that provide energy in Calories? Which three nutrients do not provide energy?

15. If you drink a beer, are you taking a drug? Explain your answer.

Concept Mapping

16. Use the following terms to create a concept map: unsaturated fats, carbohydrates, water, proteins, nutrients, simple sugars, starch, fats, vitamins, minerals, saturated fats.

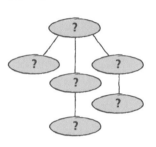

CRITICAL THINKING AND PROBLEM SOLVING

Write one or two sentences to answer the following questions:

17. Many people eat little or no meat. Meat is often the major source of protein in American diets. What other sources of protein can vegetarians choose?

18. You are at a party and a friend offers you a cigarette. He says that one cigarette won't hurt you. Using what you know about tobacco and addiction, explain why his reasoning is false.

MATH IN SCIENCE

19. Assume your diet is 20 percent fat, 30 percent protein, and 50 percent carbohydrates. In order to get 2,000 Calories a day, how many Calories of each nutrient should you be getting?

INTERPRETING GRAPHICS

List the unsafe habits in the following illustrations. For each unsafe habit, tell what the person should be doing instead.

What's wrong here?

20.

21.

Reading Check-up

Take a minute to review your answers to the Pre-Reading Questions found at the bottom of page 146. Have your answers changed? If necessary, revise your answers based on what you have learned since you began this chapter.

Science, Technology, and Society

Bacteria at Your Service

Wanted: Hard worker to clean up trash. You must have experience curing diseases. The ability to make plastics is a plus. Only microorganisms need apply for this position.

What could possibly be flexible enough to perform so many different activities? Would you believe it's bacteria?

Cleaning Up Our Act

Without bacteria, Earth would be littered with the remains of plants and animals. That's because many bacteria decompose once-living matter. Some bacteria also break down other substances. In fact, they offer solutions to some of our toughest pollution problems.

When an oil spill occurs, it often severely damages the environment. However, scientists have engineered bacteria that actually feed on oil! As the bacteria eat the oil, they break it down into harmless substances. Scientists hope to use the bacteria to clean up large oil spills, but they must first be certain that introducing a large number of these bacteria into the ocean will not harm the environment.

Producing Plastics and Pills

Did you know that some bacteria act like factories? For example, one kind of bacteria makes biodegradable plastic! These amazing bacteria store energy as plastic granules, just as animals store energy as fat and plants store energy as starch. Originally, researchers found that the plastic was stiff and brittle. However, scientists found that when certain substances are added to the bacteria's diet, the plastic they produce is flexible enough to be made into consumer products.

Other bacteria can produce important drugs used to fight disease. In fact, some bacteria even produce antibiotics that can fight infections caused by other bacteria!

Helping Plants Grow

Genetic engineers have designed bacteria that can help make plants pest resistant and cold resistant. Some bacteria even keep foods fresh longer on the grocery store shelves. From the garbage dump to the grocery store to the medicine cabinet, bacteria are at our service!

What Do You Think?

▶ Many scientists are concerned about the potential side effects of genetic engineering. What could be some consequences of altering the genetic makeup of crops? Do some research to find out for yourself. Write a report or organize a debate with your classmates.

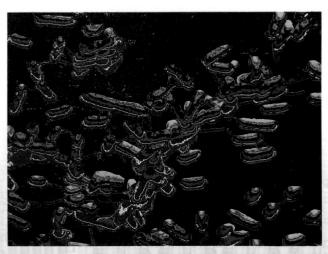

▲ *These bacteria are used to clean up oil spills.*

Meatless Munching

What'll it be today, the hamburger special, chicken surprise, or veggie platter? More and more people are opting for the veggie platter. In fact, research indicates that more than 12 million Americans are now vegetarians, and this number appears to be growing.

It's Not Just Salads

When you think of a vegetarian diet, you might think only of vegetables. Of course, vegetables are important to a vegetarian diet, but not all vegetarian diets are alike. Some vegetarian diets are based solely on plant products, while semivegetarian diets may also include dairy products, eggs, fish, and poultry.

Why the Trend?

There are many different reasons for choosing a vegetarian lifestyle. Some people believe that a vegetarian diet is healthy because a decreased consumption of animal products can lower their intake of saturated fat and cholesterol. Other people have ecological reasons for eating a vegetarian diet. For example, producing a serving of meat requires more land, water, and chemicals than producing a serving of grain. Finally, many vegetarians believe that it is unethical to kill animals for meat when plants are available.

Benefits and Risks

Recent statistics suggest that a vegetarian diet may reduce the risk of heart disease, adult-onset diabetes, and some forms of cancer. However, this type of diet takes careful thought. It is not simply a matter of eliminating meat. People may replace the meat with too

▲ *All of the foods shown here are plant products and are commonly used in vegetarian diets.*

many dairy products and eggs, which are higher in fat than most meats. Others may substitute high carbohydrate foods (such as pasta) and junk food (such as french fries) for their meat choices instead of increasing their fruit and vegetable intake. This could lead to nutritional deficiencies. The key to consuming the recommended amount of calories and nutrients for a healthy diet is to eat a wide variety of nutritious, low-fat foods. This is a good idea whether you want to decrease your meat intake or not.

Prepare a Healthy Menu

▶ Choose one of the nutrients that meats provide, and do some research to find a vegetarian substitute. What difficulties did you encounter in your search? Would you eat the substitute you found?

Exploring, inventing, and investigating are essential to the study of science. However, these activities can also be dangerous. To make sure that your experiments and explorations are safe, you must be aware of a variety of safety guidelines.

You have probably heard of the saying, "It is better to be safe than sorry." This is particularly true in a science classroom where experiments and explorations are being performed. Being uninformed and careless can result in serious injuries. Don't take chances with your own safety or with anyone else's.

Following are important guidelines for staying safe in the science classroom. Your teacher may also have safety guidelines and tips that are specific to your classroom and laboratory. Take the time to be safe.

Safety Rules!

Start Out Right

Always get your teacher's permission before attempting any laboratory exploration. Read the procedures carefully, and pay particular attention to safety information and caution statements. If you are unsure about what a safety symbol means, look it up or ask your teacher. You cannot be too careful when it comes to safety. If an accident does occur, inform your teacher immediately, regardless of how minor you think the accident is.

If you are instructed to note the odor of a substance, wave the fumes toward your nose with your hand. Never put your nose close to the source.

Safety Symbols

All of the experiments and investigations in this book and their related worksheets include important safety symbols to alert you to particular safety concerns. Become familiar with these symbols so that when you see them, you will know what they mean and what to do. It is important that you read this entire safety section to learn about specific dangers in the laboratory.

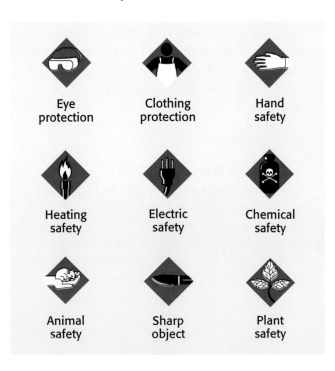

Eye protection

Clothing protection

Hand safety

Heating safety

Electric safety

Chemical safety

Animal safety

Sharp object

Plant safety

Eye Safety

Wear safety goggles when working around chemicals, acids, bases, or any type of flame or heating device. Wear safety goggles any time there is even the slightest chance that harm could come to your eyes. If any substance gets into your eyes, notify your teacher immediately, and flush your eyes with running water for at least 15 minutes. Treat any unknown chemical as if it were a dangerous chemical. Never look directly into the sun. Doing so could cause permanent blindness.

Avoid wearing contact lenses in a laboratory situation. Even if you are wearing safety goggles, chemicals can get between the contact lenses and your eyes. If your doctor requires that you wear contact lenses instead of glasses, wear eye-cup safety goggles in the lab.

Safety Equipment

Know the locations of the nearest fire alarms and any other safety equipment, such as fire blankets and eyewash fountains, as identified by your teacher, and know the procedures for using them.

Be extra careful when using any glassware. When adding a heavy object to a graduated cylinder, tilt the cylinder so the object slides slowly to the bottom.

Neatness

Keep your work area free of all unnecessary books and papers. Tie back long hair, and secure loose sleeves or other loose articles of clothing, such as ties and bows. Remove dangling jewelry. Don't wear open-toed shoes or sandals in the laboratory. Never eat, drink, or apply cosmetics in a laboratory setting. Food, drink, and cosmetics can easily become contaminated with dangerous materials.

Certain hair products (such as aerosol hair spray) are flammable and should not be worn while working near an open flame. Avoid wearing hair spray or hair gel on lab days.

Sharp/Pointed Objects

Use knives and other sharp instruments with extreme care. Never cut objects while holding them in your hands. Place objects on a suitable work surface for cutting.

Heat

Wear safety goggles when using a heating device or a flame. Whenever possible, use an electric hot plate as a heat source instead of an open flame. When heating materials in a test tube, always angle the test tube away from yourself and others. In order to avoid burns, wear heat-resistant gloves whenever instructed to do so.

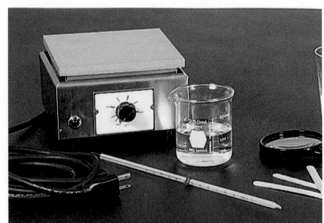

Electricity

Be careful with electrical cords. When using a microscope with a lamp, do not place the cord where it could trip someone. Do not let cords hang over a table edge in a way that could cause equipment to fall if the cord is accidentally pulled. Do not use equipment with damaged cords. Be sure your hands are dry and that the electrical equipment is in the "off" position before plugging it in. Turn off and unplug electrical equipment when you are finished.

Chemicals

Wear safety goggles when handling any potentially dangerous chemicals, acids, or bases. If a chemical is unknown, handle it as you would a dangerous chemical. Wear an apron and safety gloves when working with acids or bases or whenever you are told to do so. If a spill gets on your skin or clothing, rinse it off immediately with water for at least 5 minutes while calling to your teacher.

Never mix chemicals unless your teacher tells you to do so. Never taste, touch, or smell chemicals unless you are specifically directed to do so. Before working with a flammable liquid or gas, check for the presence of any source of flame, spark, or heat.

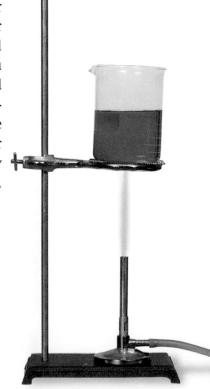

Animal Safety

Always obtain your teacher's permission before bringing any animal into the school building. Handle animals only as your teacher directs. Always treat animals carefully and with respect. Wash your hands thoroughly after handling any animal.

Plant Safety

Do not eat any part of a plant or plant seed used in the laboratory. Wash hands thoroughly after handling any part of a plant. When in nature, do not pick any wild plants unless your teacher instructs you to do so.

Glassware

Examine all glassware before use. Be sure that glassware is clean and free of chips and cracks. Report damaged glassware to your teacher. Glass containers used for heating should be made of heat-resistant glass.

Seeing Is Believing

Fingernails are part of your body's integumentary system, which includes the skin that covers your entire body. Nails are a modification of the outer layer of the skin, and they grow continuously throughout your life. In this activity, you will measure the rate at which fingernails grow.

Materials

- permanent marker
- metric ruler
- graph paper (optional)

Procedure

1. Trace around each of your hands. Then fill in some of the details, such as the fingernails. Choose a finger on your drawing, and label the parts of the fingernail, as shown at right. Notice that the nail bed is the area where the nail is attached to the finger. The illustration at right shows how far inside your finger your fingernail begins.

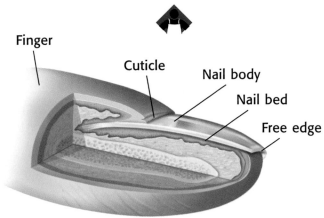

Finger
Cuticle
Nail body
Nail bed
Free edge

2. Find the center of the nail bed on your right index finger (the finger next to your thumb). Make a mark with the permanent marker on the center of your nail bed, as shown at right. **Caution:** Do not get the permanent marker ink on your clothing.

3. Measure from the mark to the base of your nail. Record this measurement on your hand drawing. Label this measurement "Day 1."

4. Repeat steps 2–3 for your left index finger. Then switch roles with your lab partner.

5. Let your fingernails grow for 2 days. Normal, daily activity will not wash away the stain completely, but you may need to freshen the mark periodically throughout this lab.

6. Measure the distance from the mark on your nail to the base of your nail. Record this distance on your hand drawing. Label this measurement "Day 3."

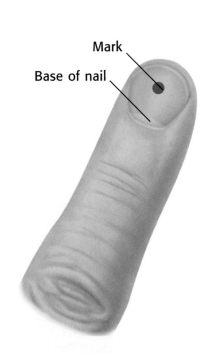

Mark
Base of nail

7. Continue measuring and recording the growth of your nails every other day for 2 weeks. Refresh the mark as necessary. You may continue to file or trim your nails as usual.

8. After you have completed your measurements, prepare a graph similar to the one below.

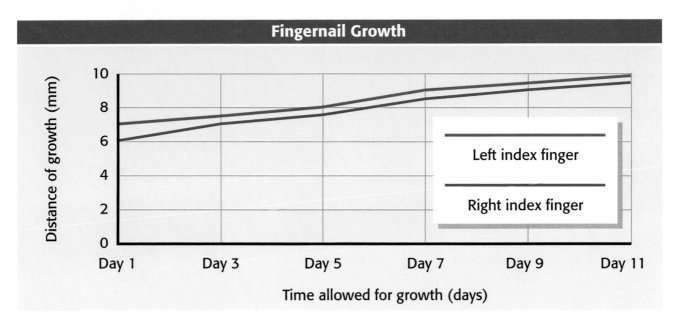

Fingernail Growth

Left index finger

Right index finger

Distance of growth (mm)

Time allowed for growth (days)

Analysis

9. Did one hand have a faster-growing nail? Write two possible explanations for this observation.

10. Who has the fastest-growing nails among your classmates? Who has the slowest-growing nails? What is the difference in the total nail growth between these two students?

11. Among your classmates, do the nail-growth rates for males and females differ? Is there a relationship between nail growth and other physical characteristics, such as height?

Going Further

Do some research in the library or on the Internet to find answers to the following questions:

• How are nails important to you? What do they help you do? Give at least three examples to support your answers.

• Are your fingernails an indication of your health or state of nutrition?

Enzymes in Action

You know how important enzymes are in the process of digestion. This lab will help you see enzymes at work. Hydrogen peroxide is continuously produced by your cells. If it is not quickly broken down, hydrogen peroxide will kill your cells. Luckily, your cells contain an enzyme that converts hydrogen peroxide into two non-poisonous substances. This enzyme is also present in the cells of beef liver. In this lab, you will observe the action of this enzyme on hydrogen peroxide.

Materials

- 1 cm cubes of beef liver (3)
- tweezers
- small plate
- 10 mL graduated cylinder
- water
- 3 test tubes
- test-tube rack
- 4 mL of fresh hydrogen peroxide
- mortar and pestle (or fork and watch glass)
- spatula
- protective gloves

Procedure

1. In your ScienceLog, draw a data table similar to the one below. Be sure to leave enough space to write your observations.

Data Table		
Size and condition of liver	**Experimental liquid**	**Observations**
1 cm cube beef liver	2 mL water	
1 cm cube beef liver	2 mL hydrogen peroxide	
1 cm cube beef liver (mashed)	2 mL hydrogen peroxide	

DO NOT WRITE IN BOOK

2. Get three equal-sized pieces of beef liver from your teacher, and use your forceps to place them on your plate.

3. Pour 2 mL of water into a test tube labeled "Water and liver."

4. Using the tweezers, carefully place one piece of liver in the test tube. Record your observations in your data table.

5. Pour 2 mL of hydrogen peroxide into a second test tube labeled "Liver and hydrogen peroxide."
   Caution: Do not splash hydrogen peroxide on your skin. If you do get hydrogen peroxide on your skin, rinse the affected area with running water immediately, and tell your teacher.

6. Using the tweezers, carefully place one piece of liver in the test tube. Record your observations of the second test tube in your data table.

7. Pour another 2 mL of hydrogen peroxide into a third test tube labeled "Ground liver and hydrogen peroxide."

8. Using a mortar and pestle (or fork and watch glass), carefully grind the third piece of liver.

9. Using the spatula, scrape the ground liver into the third test tube. Record your observations of the third test tube in your data table.

Analysis

10. What was the purpose of putting the first piece of liver in water? Why was this a necessary step?

11. Describe the difference you observed between the liver and the ground liver when each was placed in the hydrogen peroxide. How can you account for this difference?

Going Further
Do plant cells contain enzymes that break down hydrogen peroxide? Try this experiment using potato cubes instead of liver to find out.

My, How You've Grown!

In humans, the process of development that takes place between fertilization and birth lasts about 266 days. In 4 weeks, the new individual grows from a single fertilized cell to an embryo whose heart is beating and pumping blood. All of the organ systems and body parts are completely formed by the end of the seventh month. During the last 2 months before birth, the baby grows and its organ systems mature. At birth, the average mass of a baby is about 33,000 times as much as that of an embryo at 2 weeks of development! In this activity you will discover just how fast a fetus grows.

SKILL BUILDER

Materials

- graph paper
- colored pencils

TRY at HOME

Procedure

1. Using graph paper, make two graphs—one titled "Length" and one titled "Mass"—in your ScienceLog. On the length graph, use intervals of 25 mm on the *y*-axis. Extend the *y*-axis to 500 mm. On the mass graph, use intervals of 100 g on the *y*-axis. Extend this *y*-axis to 3,300 g. Use 2-week intervals for time on the *x*-axes for both graphs. Both *x*-axes should extend to 40 weeks.

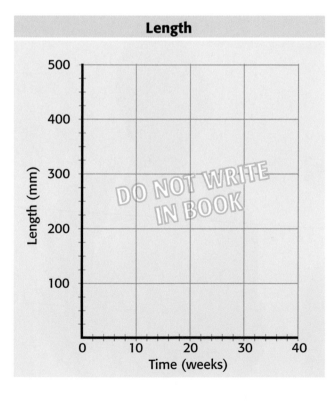

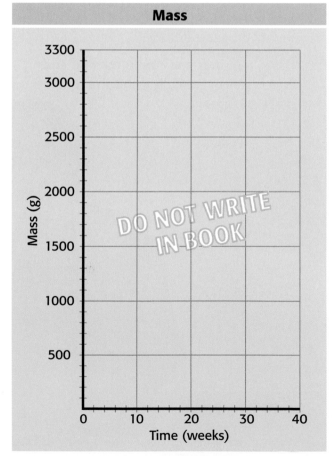

2. Examine the data table at right. Plot the data in the table on your graphs. Use a colored pencil to draw the curved line that joins the points on each graph.

Analysis

3. Describe the change in mass of a developing fetus. How can you explain this change?

4. Describe the change in length of a developing fetus. How does the change in mass compare to the change in length?

Going Further

Using the information in your graphs, estimate how tall a child would be at age 3 if he or she continued to grow at the same average rate a fetus grows.

Increase of Mass and Length of Average Human Fetus		
Time (wks)	Mass (g)	Length (mm)
2	0.1	1.5
3	0.3	2.3
4	0.5	5.0
5	0.6	10.0
6	0.8	15.0
8	1.0	30.0
13	15.0	90.0
17	115.0	140.0
21	300.0	250.0
26	950.0	320.0
30	1,500.0	400.0
35	2,300.0	450.0
40	3,300.0	500.0

DISCOVERY LAB

Passing the Cold

There are more than 100 viruses that cause the symptoms of the common cold. Any of the viruses can be passed from person to person—through the air or through direct contact. In this activity you will track the progress of an outbreak in your class.

Materials

- 200 mL beaker or a cup of similar size
- eyedropper
- 50 mL of an unknown solution
- protective gloves

Ask a Question

1. How are cold viruses passed from person to person? How can the progress of an outbreak be modeled?

Conduct an Experiment

2. Obtain an empty cup or beaker, an eyedropper, and 50 mL of one of the solutions from your teacher. Only one of you will have the "cold virus" solution. You will see a change in your solution when you have become infected.

3. Your teacher will divide the class into two equal groups. If there is an extra student, that person will record data on the chalkboard. Otherwise, the teacher will act as the recorder.

4. Each group will form a straight line and face each other.

5. Each time your teacher says "Mix," fill your eyedropper with your solution, and place 10 drops of your solution in the beaker of the person in the line opposite you without touching your eyedropper to the liquid.

6. Gently stir the liquid in your cup with your eyedropper. Do not put your eyedropper in anyone else's solution.

7. If your solution changes color, raise your hand so that the recorder can record the number of students who have been "infected."

8. Your teacher will instruct one line to move one person to the right. The person at the end of the line without a partner should go to the other end of the line.

9. Repeat steps 5–8 nine more times for a total of 10 trials.

Collect Data

10. Return to your desk, and create a data table in your ScienceLog similar to the one below. The column with the title "Total number of people" will remain the same in every row. Enter the data on the board into your data table.

11. Find the percentage of infected people for the last column by dividing the number of infected people by the total number of people and multiplying by 100 in each line.

	Results		
Trial	Number of infected people	Total number of people	Percentage of infected people
1			
2			
3			
4			
5			
6			
7			
8			
9			
10			

Analyze the Results

12. Did you become infected? If so, during which trial did you become infected?

13. Did everyone eventually become infected? If so, how many trials were necessary to infect everyone?

14. Explain at least one reason why this simulation may underestimate the number of people who might have been infected in real life.

15. Use your results to create a line graph showing the change in the infection percentage per trial.

Going Further

Research in the library or on the Internet to find out some of the factors that contribute to the spread of a cold virus. What is the best and easiest way to reduce your chances of catching a cold? Explain your answer.

Concept Mapping: A Way to Bring Ideas Together

What Is a Concept Map?

Have you ever tried to tell someone about a book or a chapter you've just read and found that you can remember only a few isolated words and ideas? Or maybe you've memorized facts for a test and then weeks later discovered you're not even sure what topics those facts covered.

In both cases, you may have understood the ideas or concepts by themselves but not in relation to one another. If you could somehow link the ideas together, you would probably understand them better and remember them longer. This is something a concept map can help you do. A concept map is a way to see how ideas or concepts fit together. It can help you see the "big picture."

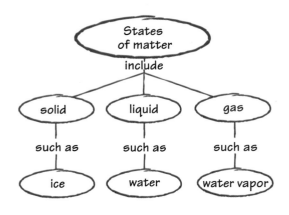

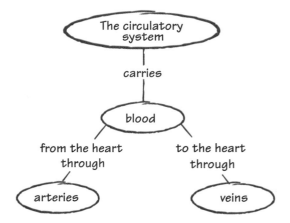

How to Make a Concept Map

❶ Make a list of the main ideas or concepts.

It might help to write each concept on its own slip of paper. This will make it easier to rearrange the concepts as many times as necessary to make sense of how the concepts are connected. After you've made a few concept maps this way, you can go directly from writing your list to actually making the map.

❷ Arrange the concepts in order from the most general to the most specific.

Put the most general concept at the top and circle it. Ask yourself, "How does this concept relate to the remaining concepts?" As you see the relationships, arrange the concepts in order from general to specific.

❸ Connect the related concepts with lines.

❹ On each line, write an action word or short phrase that shows how the concepts are related.

Look at the concept maps on this page, and then see if you can make one for the following terms:

plants, water, photosynthesis, carbon dioxide, sun's energy

One possible answer is provided at right, but don't look at it until you try the concept map yourself.

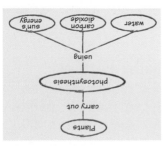

SI Measurement

The International System of Units, or SI, is the standard system of measurement used by many scientists. Using the same standards of measurement makes it easier for scientists to communicate with one another.

SI works by combining prefixes and base units. Each base unit can be used with different prefixes to define smaller and larger quantities. The table below lists common SI prefixes.

SI Prefixes			
Prefix	Abbreviation	Factor	Example
kilo-	k	1,000	kilogram, 1 kg = 1,000 g
hecto-	h	100	hectoliter, 1 hL = 100 L
deka-	da	10	dekameter, 1 dam = 10 m
		1	meter, liter
deci-	d	0.1	decigram, 1 dg = 0.1 g
centi-	c	0.01	centimeter, 1 cm = 0.01 m
milli-	m	0.001	milliliter, 1 mL = 0.001 L
micro-	μ	0.000 001	micrometer, 1 μm = 0.000 001 m

SI Conversion Table		
SI units	From SI to English	From English to SI
Length		
kilometer (km) = 1,000 m	1 km = 0.621 mi	1 mi = 1.609 km
meter (m) = 100 cm	1 m = 3.281 ft	1 ft = 0.305 m
centimeter (cm) = 0.01 m	1 cm = 0.394 in.	1 in. = 2.540 cm
millimeter (mm) = 0.001 m	1 mm = 0.039 in.	
micrometer (μm) = 0.000 001 m		
nanometer (nm) = 0.000 000 001 m		
Area		
square kilometer (km^2) = 100 hectares	1 km^2 = 0.386 mi^2	1 mi^2 = 2.590 km^2
hectare (ha) = 10,000 m^2	1 ha = 2.471 acres	1 acre = 0.405 ha
square meter (m^2) = 10,000 cm^2	1 m^2 = 10.765 ft^2	1 ft^2 = 0.093 m^2
square centimeter (cm^2) = 100 mm^2	1 cm^2 = 0.155 in.^{2}	1 in.^{2} = 6.452 cm^2
Volume		
liter (L) = 1,000 mL = 1 dm^3	1 L = 1.057 fl qt	1 fl qt = 0.946 L
milliliter (mL) = 0.001 L = 1 cm^3	1 mL = 0.034 fl oz	1 fl oz = 29.575 mL
microliter (μL) = 0.000 001 L		
Mass		
kilogram (kg) = 1,000 g	1 kg = 2.205 lb	1 lb = 0.454 kg
gram (g) = 1,000 mg	1 g = 0.035 oz	1 oz = 28.349 g
milligram (mg) = 0.001 g		
microgram (μg) = 0.000 001 g		

Temperature Scales

Temperature can be expressed using three different scales: Fahrenheit, Celsius, and Kelvin. The SI unit for temperature is the kelvin (K).

Although 0 K is much colder than 0°C, a change of 1 K is equal to a change of 1°C.

Three Temperature Scales			
	Fahrenheit	**Celsius**	**Kelvin**
Water boils	212°	100°	373
Body temperature	98.6°	37°	310
Room temperature	68°	20°	293
Water freezes	32°	0°	273

Temperature Conversions Table		
To convert	**Use this equation:**	**Example**
Celsius to Fahrenheit °C ⟶ °F	$°F = \left(\dfrac{9}{5} \times °C\right) + 32$	Convert 45°C to °F. $°F = \left(\dfrac{9}{5} \times 45°C\right) + 32 = 113°F$
Fahrenheit to Celsius °F ⟶ °C	$°C = \dfrac{5}{9} \times (°F - 32)$	Convert 68°F to °C. $°C = \dfrac{5}{9} \times (68°F - 32) = 20°C$
Celsius to Kelvin °C ⟶ K	$K = °C + 273$	Convert 45°C to K. $K = 45°C + 273 = 318\ K$
Kelvin to Celsius K ⟶ °C	$°C = K - 273$	Convert 32 K to °C. $°C = 32\ K - 273 = -241°C$

Measuring Skills

Using a Graduated Cylinder

When using a graduated cylinder to measure volume, keep the following procedures in mind:

1 Make sure the cylinder is on a flat, level surface.

2 Move your head so that your eye is level with the surface of the liquid.

3 Read the mark closest to the liquid level. On glass graduated cylinders, read the mark closest to the center of the curve in the liquid's surface.

Using a Meterstick or Metric Ruler

When using a meterstick or metric ruler to measure length, keep the following procedures in mind:

1 Place the ruler firmly against the object you are measuring.

2 Align one edge of the object exactly with the zero end of the ruler.

3 Look at the other edge of the object to see which of the marks on the ruler is closest to that edge. **Note:** Each small slash between the centimeters represents a millimeter, which is one-tenth of a centimeter.

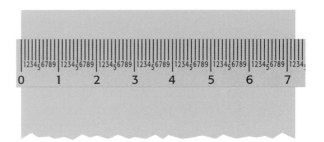

Using a Triple-Beam Balance

When using a triple-beam balance to measure mass, keep the following procedures in mind:

1 Make sure the balance is on a level surface.

2 Place all of the countermasses at zero. Adjust the balancing knob until the pointer rests at zero.

3 Place the object you wish to measure on the pan. **Caution:** Do not place hot objects or chemicals directly on the balance pan.

4 Move the largest countermass along the beam to the right until it is at the last notch that does not tip the balance. Follow the same procedure with the next-largest countermass. Then move the smallest countermass until the pointer rests at zero.

5 Add the readings from the three beams together to determine the mass of the object.

6 When determining the mass of crystals or powders, use a piece of filter paper. First find the mass of the paper. Then add the crystals or powder to the paper and re-measure. The actual mass of the crystals or powder is the total mass minus the mass of the paper. When finding the mass of liquids, first find the mass of the empty container. Then find the mass of the liquid and container together. The mass of the liquid is the total mass minus the mass of the container.

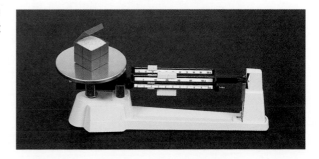

Scientific Method

The series of steps that scientists use to answer questions and solve problems is often called the **scientific method.** The scientific method is not a rigid procedure. Scientists may use all of the steps or just some of the steps of the scientific method. They may even repeat some of the steps. The goal of the scientific method is to come up with reliable answers and solutions.

Six Steps of the Scientific Method

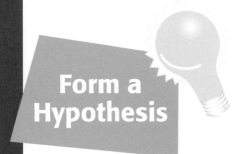

1 **Ask a Question** Good questions come from careful **observations.** You make observations by using your senses to gather information. Sometimes you may use instruments, such as microscopes and telescopes, to extend the range of your senses. As you observe the natural world, you will discover that you have many more questions than answers. These questions drive the scientific method.

Questions beginning with *what, why, how,* and *when* are very important in focusing an investigation, and they often lead to a hypothesis. (You will learn what a hypothesis is in the next step.) Here is an example of a question that could lead to further investigation.

Question: How does acid rain affect plant growth?

2 **Form a Hypothesis** After you come up with a question, you need to turn the question into a **hypothesis.** A hypothesis is a clear statement of what you expect the answer to your question to be. Your hypothesis will represent your best "educated guess" based on your observations and what you already know. A good hypothesis is testable. If observations and information cannot be gathered or if an experiment cannot be designed to test your hypothesis, it is untestable, and the investigation can go no further.

Here is a hypothesis that could be formed from the question, "How does acid rain affect plant growth?"

Hypothesis: Acid rain causes plants to grow more slowly.

Notice that the hypothesis provides some specifics that lead to methods of testing. The hypothesis can also lead to predictions. A **prediction** is what you think will be the outcome of your experiment or data collection. Predictions are usually stated in an "if . . . then" format. For example, **if** meat is kept at room temperature, **then** it will spoil faster than meat kept in the refrigerator. More than one prediction can be made for a single hypothesis. Here is a sample prediction for the hypothesis that acid rain causes plants to grow more slowly.

Prediction: If a plant is watered with only acid rain (which has a pH of 4), then the plant will grow at half its normal rate.

3 **Test the Hypothesis** After you have formed a hypothesis and made a prediction, you should test your hypothesis. There are different ways to do this. Perhaps the most familiar way is to conduct a **controlled experiment.** A controlled experiment tests only one factor at a time. A controlled experiment has a **control group** and one or more **experimental groups.** All the factors for the control and experimental groups are the same except for one factor, which is called the **variable.** By changing only one factor, you can see the results of just that one change.

Test the Hypothesis

Sometimes, the nature of an investigation makes a controlled experiment impossible. For example, dinosaurs have been extinct for millions of years, and the Earth's core is surrounded by thousands of meters of rock. It would be difficult, if not impossible, to conduct controlled experiments on such things. Under such circumstances, a hypothesis may be tested by making detailed observations. Taking measurements is one way of making observations.

4 **Analyze the Results** After you have completed your experiments, made your observations, and collected your data, you must analyze all the information you have gathered. Tables and graphs are often used in this step to organize the data.

Analyze the Results

5 **Draw Conclusions** Based on the analysis of your data, you should conclude whether or not your results support your hypothesis. If your hypothesis is supported, you (or others) might want to repeat the observations or experiments to verify your results. If your hypothesis is not supported by the data, you may have to check your procedure for errors. You may even have to reject your hypothesis and make a new one. If you cannot draw a conclusion from your results, you may have to try the investigation again or carry out further observations or experiments.

Draw Conclusions

Do they support your hypothesis?

No

Yes

6 **Communicate Results** After any scientific investigation, you should report your results. By doing a written or oral report, you let others know what you have learned. They may want to repeat your investigation to see if they get the same results. Your report may even lead to another question, which in turn may lead to another investigation.

Communicate Results

Scientific Method in Action

The scientific method is not a "straight line" of steps. It contains loops in which several steps may be repeated over and over again, while others may not be necessary. For example, sometimes scientists will find that testing one hypothesis raises new questions and new hypotheses to be tested. And sometimes, testing the hypothesis leads directly to a conclusion. Furthermore, the steps in the scientific method are not always used in the same order. Follow the steps in the diagram below, and see how many different directions the scientific method can take you.

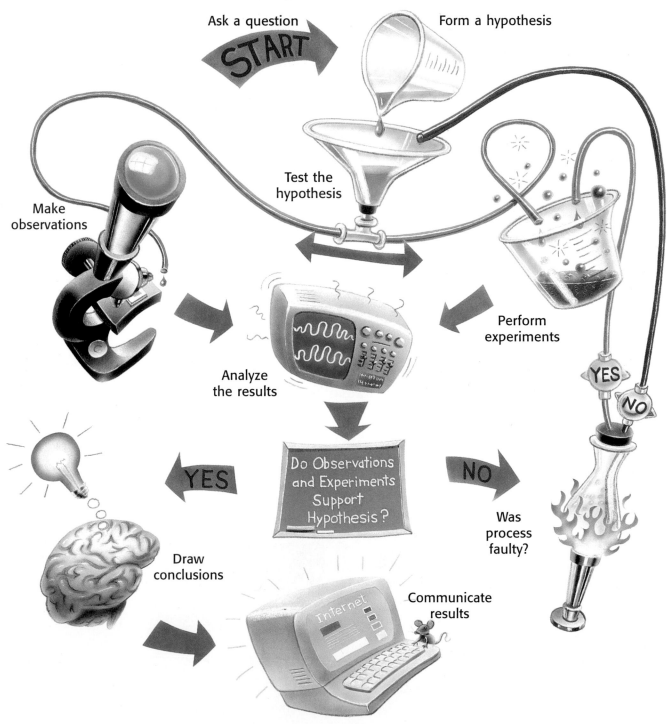

Ask a question

START

Form a hypothesis

Make observations

Test the hypothesis

Perform experiments

Analyze the results

YES

NO

YES

Do Observations and Experiments Support Hypothesis?

NO

Was process faulty?

Draw conclusions

Communicate results

Internet

Making Charts and Graphs

Circle Graphs

A circle graph, or pie chart, shows how each group of data relates to all of the data. Each part of the circle represents a category of the data. The entire circle represents all of the data. For example, a biologist studying a hardwood forest in Wisconsin found that there were five different types of trees. The data table at right summarizes the biologist's findings.

Wisconsin Hardwood Trees	
Type of tree	Number found
Oak	600
Maple	750
Beech	300
Birch	1,200
Hickory	150
Total	3,000

How to Make a Circle Graph

1 In order to make a circle graph of this data, first find the percentage of each type of tree. To do this, divide the number of individual trees by the total number of trees and multiply by 100.

$$\frac{600 \text{ oak}}{3,000 \text{ trees}} \times 100 = 20\%$$

$$\frac{750 \text{ maple}}{3,000 \text{ trees}} \times 100 = 25\%$$

$$\frac{300 \text{ beech}}{3,000 \text{ trees}} \times 100 = 10\%$$

$$\frac{1,200 \text{ birch}}{3,000 \text{ trees}} \times 100 = 40\%$$

$$\frac{150 \text{ hickory}}{3,000 \text{ trees}} \times 100 = 5\%$$

2 Now determine the size of the pie shapes that make up the chart. Do this by multiplying each percentage by 360°. Remember that a circle contains 360°.

20% × 360° = 72°	25% × 360° = 90°
10% × 360° = 36°	40% × 360° = 144°
5% × 360° = 18°	

3 Then check that the sum of the percentages is 100 and the sum of the degrees is 360.

20% + 25% + 10% + 40% + 5% = 100%
72° + 90° + 36° + 144° + 18° = 360°

4 Use a compass to draw a circle and mark its center.

5 Then use a protractor to draw angles of 72°, 90°, 36°, 144°, and 18° in the circle.

6 Finally, label each part of the graph, and choose an appropriate title.

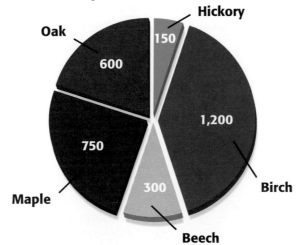

A Community of Wisconsin Hardwood Trees

Line Graphs

Line graphs are most often used to demonstrate continuous change. For example, Mr. Smith's science class analyzed the population records for their hometown, Appleton, between 1900 and 2000. Examine the data at left.

Because the year and the population change, they are the *variables*. The population is determined by, or dependent on, the year. Therefore, the population is called the **dependent variable**, and the year is called the **independent variable**. Each set of data is called a **data pair**. To prepare a line graph, data pairs must first be organized in a table like the one at left.

Population of Appleton, 1900–2000	
Year	Population
1900	1,800
1920	2,500
1940	3,200
1960	3,900
1980	4,600
2000	5,300

How to Make a Line Graph

1 Place the independent variable along the horizontal (*x*) axis. Place the dependent variable along the vertical (*y*) axis.

2 Label the *x*-axis "Year" and the *y*-axis "Population." Look at your largest and smallest values for the population. Determine a scale for the *y*-axis that will provide enough space to show these values. You must use the same scale for the entire length of the axis. Find an appropriate scale for the *x*-axis too.

3 Choose reasonable starting points for each axis.

4 Plot the data pairs as accurately as possible.

5 Choose a title that accurately represents the data.

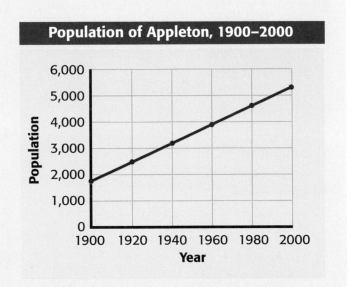

Population of Appleton, 1900–2000

How to Determine Slope

Slope is the ratio of the change in the *y*-axis to the change in the *x*-axis, or "rise over run."

1 Choose two points on the line graph. For example, the population of Appleton in 2000 was 5,300 people. Therefore, you can define point *a* as (2000, 5,300). In 1900, the population was 1,800 people. Define point *b* as (1900, 1,800).

2 Find the change in the *y*-axis.
(*y* at point *a*) − (*y* at point *b*)
5,300 people − 1,800 people = 3,500 people

3 Find the change in the *x*-axis.
(*x* at point *a*) − (*x* at point *b*)
2000 − 1900 = 100 years

4 Calculate the slope of the graph by dividing the change in *y* by the change in *x*.

$$\text{slope} = \frac{\text{change in } y}{\text{change in } x}$$

$$\text{slope} = \frac{3{,}500 \text{ people}}{100 \text{ years}}$$

slope = 35 people per year

In this example, the population in Appleton increased by a fixed amount each year. The graph of this data is a straight line. Therefore, the relationship is **linear**. When the graph of a set of data is not a straight line, the relationship is **nonlinear**.

Using Algebra to Determine Slope

The equation in step 4 may also be arranged to be:

$$y = kx$$

where y represents the change in the y-axis, k represents the slope, and x represents the change in the x-axis.

$$\text{slope} = \frac{\text{change in } y}{\text{change in } x}$$

$$k = \frac{y}{x}$$

$$k \times x = \frac{y \times x}{x}$$

$$kx = y$$

Bar Graphs

Bar graphs are used to demonstrate change that is not continuous. These graphs can be used to indicate trends when the data are taken over a long period of time. A meteorologist gathered the precipitation records at right for Hartford, Connecticut, for April 1–15, 1996, and used a bar graph to represent the data.

Precipitation in Hartford, Connecticut April 1–15, 1996

Date	Precipitation (cm)	Date	Precipitation (cm)
April 1	0.5	April 9	0.25
April 2	1.25	April 10	0.0
April 3	0.0	April 11	1.0
April 4	0.0	April 12	0.0
April 5	0.0	April 13	0.25
April 6	0.0	April 14	0.0
April 7	0.0	April 15	6.50
April 8	1.75		

How to Make a Bar Graph

❶ Use an appropriate scale and a reasonable starting point for each axis.

❷ Label the axes, and plot the data.

❸ Choose a title that accurately represents the data.

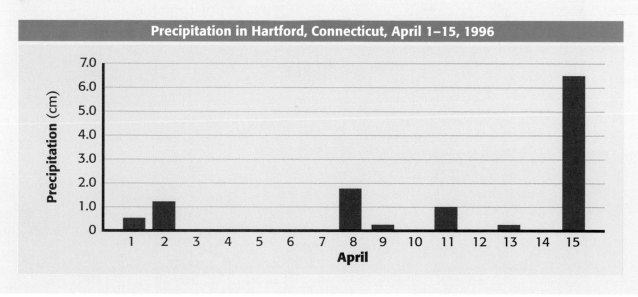

Precipitation in Hartford, Connecticut, April 1–15, 1996

Periodic Table of the Elements

Each square on the table includes an element's name, chemical symbol, atomic number, and atomic mass.

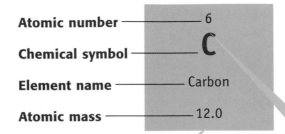

Atomic number —— 6

Chemical symbol —— C

Element name —— Carbon

Atomic mass —— 12.0

The background color indicates the type of element. Carbon is a nonmetal.

The color of the chemical symbol indicates the physical state at room temperature. Carbon is a solid.

Background
- Metals
- Metalloids
- Nonmetals

Chemical Symbol
- Solid
- Liquid
- Gas

Period 1

1
H
Hydrogen
1.0

	Group 1	Group 2
Period 2	3 **Li** Lithium 6.9	4 **Be** Beryllium 9.0
Period 3	11 **Na** Sodium 23.0	12 **Mg** Magnesium 24.3

	Group 1	Group 2	Group 3	Group 4	Group 5	Group 6	Group 7	Group 8	Group 9
Period 4	19 **K** Potassium 39.1	20 **Ca** Calcium 40.1	21 **Sc** Scandium 45.0	22 **Ti** Titanium 47.9	23 **V** Vanadium 50.9	24 **Cr** Chromium 52.0	25 **Mn** Manganese 54.9	26 **Fe** Iron 55.8	27 **Co** Cobalt 58.9
Period 5	37 **Rb** Rubidium 85.5	38 **Sr** Strontium 87.6	39 **Y** Yttrium 88.9	40 **Zr** Zirconium 91.2	41 **Nb** Niobium 92.9	42 **Mo** Molybdenum 95.9	43 **Tc** Technetium (97.9)	44 **Ru** Ruthenium 101.1	45 **Rh** Rhodium 102.9
Period 6	55 **Cs** Cesium 132.9	56 **Ba** Barium 137.3	57 **La** Lanthanum 138.9	72 **Hf** Hafnium 178.5	73 **Ta** Tantalum 180.9	74 **W** Tungsten 183.8	75 **Re** Rhenium 186.2	76 **Os** Osmium 190.2	77 **Ir** Iridium 192.2
Period 7	87 **Fr** Francium (223.0)	88 **Ra** Radium (226.0)	89 **Ac** Actinium (227.0)	104 **Rf** Rutherfordium (261.1)	105 **Db** Dubnium (262.1)	106 **Sg** Seaborgium (263.1)	107 **Bh** Bohrium (262.1)	108 **Hs** Hassium (265)	109 **Mt** Meitnerium (266)

A row of elements is called a period.

A column of elements is called a group or family.

Lanthanides	58 **Ce** Cerium 140.1	59 **Pr** Praseodymium 140.9	60 **Nd** Neodymium 144.2	61 **Pm** Promethium (144.9)	62 **Sm** Samarium 150.4
Actinides	90 **Th** Thorium 232.0	91 **Pa** Protactinium 231.0	92 **U** Uranium 238.0	93 **Np** Neptunium (237.0)	94 **Pu** Plutonium 244.1

These elements are placed below the table to allow the table to be narrower.

APPENDIX

This zigzag line reminds you where the metals, nonmetals, and metalloids are.

Group 18

Group 13	Group 14	Group 15	Group 16	Group 17	2 He Helium 4.0
5 B Boron 10.8	6 C Carbon 12.0	7 N Nitrogen 14.0	8 O Oxygen 16.0	9 F Fluorine 19.0	10 Ne Neon 20.2

Group 10	Group 11	Group 12	13 Al Aluminum 27.0	14 Si Silicon 28.1	15 P Phosphorus 31.0	16 S Sulfur 32.1	17 Cl Chlorine 35.5	18 Ar Argon 39.9
28 Ni Nickel 58.7	29 Cu Copper 63.5	30 Zn Zinc 65.4	31 Ga Gallium 69.7	32 Ge Germanium 72.6	33 As Arsenic 74.9	34 Se Selenium 79.0	35 Br Bromine 79.9	36 Kr Krypton 83.8
46 Pd Palladium 106.4	47 Ag Silver 107.9	48 Cd Cadmium 112.4	49 In Indium 114.8	50 Sn Tin 118.7	51 Sb Antimony 121.8	52 Te Tellurium 127.6	53 I Iodine 126.9	54 Xe Xenon 131.3
78 Pt Platinum 195.1	79 Au Gold 197.0	80 Hg Mercury 200.6	81 Tl Thallium 204.4	82 Pb Lead 207.2	83 Bi Bismuth 209.0	84 Po Polonium (209.0)	85 At Astatine (210.0)	86 Rn Radon (222.0)
110 Uun* Ununnilium (271)	111 Uuu* Unununium (272)	112 Uub* Ununbium (277)		114 Uuq* Ununquadium (285)		116 Uuh* Ununhexium (289)		118 Uuo* Ununoctium (293)

A number in parenthesis is the mass number of the most stable form of that element.

63 Eu Europium 152.0	64 Gd Gadolinium 157.3	65 Tb Terbium 158.9	66 Dy Dysprosium 162.5	67 Ho Holmium 164.9	68 Er Erbium 167.3	69 Tm Thulium 168.9	70 Yb Ytterbium 173.0	71 Lu Lutetium 175.0
95 Am Americium (243.1)	96 Cm Curium (247.1)	97 Bk Berkelium (247.1)	98 Cf Californium (251.1)	99 Es Einsteinium (252.1)	100 Fm Fermium (257.1)	101 Md Mendelevium (258.1)	102 No Nobelium (259.1)	103 Lr Lawrencium (262.1)

*The official names and symbols for the elements greater than 109 will eventually be approved by a committee of scientists.

The Six Kingdoms

Kingdom Archaebacteria

The organisms in this kingdom are single-celled prokaryotes.

Archaebacteria		
Group	**Examples**	**Characteristics**
Methanogens	*Methanococcus*	found in soil, swamps, the digestive tract of mammals; produce methane gas; can't live in oxygen
Thermophiles	*Sulpholobus*	found in extremely hot environments; require sulphur, can't live in oxygen
Halophiles	*Halococcus*	found in environments with very high salt content, such as the Dead Sea; nearly all can live in oxygen

Kingdom Eubacteria

There are more than 4,000 named species in this kingdom of single-celled prokaryotes.

Eubacteria		
Group	**Examples**	**Characteristics**
Bacilli	*Escherichia coli*	rod-shaped; free-living, symbiotic, or parasitic; some can fix nitrogen; some cause disease
Cocci	*Streptococcus*	spherical-shaped, disease-causing; can form spores to resist unfavorable environments
Spirilla	*Treponema*	spiral-shaped; responsible for several serious illnesses, such as syphilis and Lyme disease

Kingdom Protista

The organisms in this kingdom are eukaryotes. There are single-celled and multicellular representatives.

Protists		
Group	**Examples**	**Characteristics**
Sacodines	*Amoeba*	radiolarians; single-celled consumers
Ciliates	*Paramecium*	single-celled consumers
Flagellates	*Trypanosoma*	single-celled parasites
Sporozoans	*Plasmodium*	single-celled parasites
Euglenas	*Euglena*	single-celled; photosynthesize
Diatoms	*Pinnularia*	most are single-celled; photosynthesize
Dinoflagellates	*Gymnodinium*	single-celled; some photosynthesize
Algae	*Volvox*, coral algae	4 phyla; single- or many-celled; photosynthesize
Slime molds	*Physarum*	single- or many-celled; consumers or decomposers
Water molds	powdery mildew	single- or many-celled, parasites or decomposers

Kingdom Fungi

There are single-celled and multicellular eukaryotes in this kingdom. There are four major groups of fungi.

Fungi		
Group	**Examples**	**Characteristics**
Threadlike fungi	bread mold	spherical; decomposers
Sac fungi	yeast, morels	saclike; parasites and decomposers
Club fungi	mushrooms, rusts, smuts	club-shaped; parasites and decomposers
Lichens	British soldier	symbiotic with algae

Kingdom Plantae

The organisms in this kingdom are multicellular eukaryotes. They have specialized organ systems for different life processes. They are classified in divisions instead of phyla.

Plants		
Group	**Examples**	**Characteristics**
Bryophytes	mosses, liverworts	reproduce by spores
Club mosses	*Lycopodium*, ground pine	reproduce by spores
Horsetails	rushes	reproduce by spores
Ferns	spleenworts, sensitive fern	reproduce by spores
Conifers	pines, spruces, firs	reproduce by seeds; cones
Cycads	*Zamia*	reproduce by seeds
Gnetophytes	*Welwitschia*	reproduce by seeds
Ginkgoes	*Ginkgo*	reproduce by seeds
Angiosperms	all flowering plants	reproduce by seeds; flowers

Kingdom Animalia

This kingdom contains multicellular eukaryotes. They have specialized tissues and complex organ systems.

Animals		
Group	**Examples**	**Characteristics**
Sponges	glass sponges	no symmetry or segmentation; aquatic
Cnidarians	jellyfish, coral	radial symmetry; aquatic
Flatworms	planaria, tapeworms, flukes	bilateral symmetry; organ systems
Roundworms	*Trichina*, hookworms	bilateral symmetry; organ systems
Annelids	earthworms, leeches	bilateral symmetry; organ systems
Mollusks	snails, octopuses	bilateral symmetry; organ systems
Echinoderms	sea stars, sand dollars	radial symmetry; organ systems
Arthropods	insects, spiders, lobsters	bilateral symmetry; organ systems
Chordates	fish, amphibians, reptiles, birds, mammals	bilateral symmetry; complex organ systems

Using the Microscope

Parts of the Compound Light Microscope

- The **ocular lens** magnifies the image 10×.
- The **low-power objective** magnifies the image 10×.
- The **high-power objective** magnifies the image either 40× or 43×.
- The **revolving nosepiece** holds the objectives and can be turned to change from one magnification to the other.
- The **body tube** maintains the correct distance between the ocular lens and objectives.
- The **coarse-adjustment knob** moves the body tube up and down to allow focusing of the image.
- The **fine-adjustment knob** moves the body tube slightly to bring the image into sharper focus.
- The **stage** supports a slide.
- **Stage clips** hold the slide in place for viewing.
- The **diaphragm** controls the amount of light coming through the stage.
- The light source provides a **light** for viewing the slide.
- The **arm** supports the body tube.
- The **base** supports the microscope.

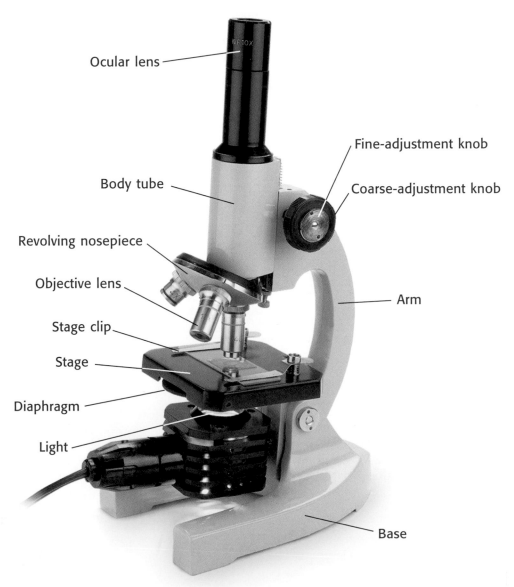

Ocular lens

Fine-adjustment knob

Body tube

Coarse-adjustment knob

Revolving nosepiece

Objective lens

Arm

Stage clip

Stage

Diaphragm

Light

Base

Proper Use of the Compound Light Microscope

1 Carry the microscope to your lab table using both hands. Place one hand beneath the base, and use the other hand to hold the arm of the microscope. Hold the microscope close to your body while moving it to your lab table.

2 Place the microscope on the lab table at least 5 cm from the edge of the table.

3 Check to see what type of light source is used by your microscope. If the microscope has a lamp, plug it in, making sure that the cord is out of the way. If the microscope has a mirror, adjust it to reflect light through the hole in the stage.
Caution: If your microscope has a mirror, do not use direct sunlight as a light source. Direct sunlight can damage your eyes.

4 Always begin work with the low-power objective in line with the body tube. Adjust the revolving nosepiece.

5 Place a prepared slide over the hole in the stage. Secure the slide with the stage clips.

6 Look through the ocular lens. Move the diaphragm to adjust the amount of light coming through the stage.

7 Look at the stage from eye level. Slowly turn the coarse adjustment to lower the objective until it almost touches the slide. Do not allow the objective to touch the slide.

8 Look through the ocular lens. Turn the coarse adjustment to raise the low-power objective until the image is in focus. Always focus by raising the objective away from the slide. *Never focus the objective downward.* Use the fine adjustment to sharpen the focus. Keep both eyes open while viewing a slide.

9 Make sure that the image is exactly in the center of your field of vision. Then switch to the high-power objective. Focus the image, using only the fine adjustment. *Never use the coarse adjustment at high power.*

10 When you are finished using the microscope, remove the slide. Clean the ocular lens and objective lenses with lens paper. Return the microscope to its storage area. Remember, you should use both hands to carry the microscope.

Making a Wet Mount

1 Use lens paper to clean a glass slide and a coverslip.

2 Place the specimen you wish to observe in the center of the slide.

3 Using a medicine dropper, place one drop of water on the specimen.

4 Hold the coverslip at the edge of the water and at a 45° angle to the slide. Make sure that the water runs along the edge of the coverslip.

5 Lower the coverslip slowly to avoid trapping air bubbles.

6 Water might evaporate from the slide as you work. Add more water to keep the specimen fresh. Place the tip of the medicine dropper next to the edge of the coverslip. Add a drop of water. (You can also use this method to add stain or solutions to a wet mount.) Remove excess water from the slide by using the corner of a paper towel as a blotter. Do not lift the coverslip to add or remove water.

Glossary

A

addiction   a physical dependence on a drug (156)

aerobic exercise   vigorous, sustained exercise of the whole body for 20 minutes or more (162)

alcoholism   a disorder in which a person is physically and psychologically dependent on alcohol (158)

allergy   an inappropriate immune-system reaction to a harmless antigen (135)

alveoli   (al VEE uh LIE) tiny sacs that form the bronchiole branches of the lungs (41)

amnion   a thin, fluid-filled membrane surrounding a placental mammal's fetus (111)

anorexia nervosa   a disorder characterized by self-starvation and an intense fear of gaining weight (154)

antibiotic   a substance used to kill or slow the growth of bacteria or other microorganisms (129)

antibody   a special protein that can recognize specific pathogens (131)

antigen   pieces of a pathogen that generate an immune response from immune-system cells (132)

anus   the opening at the end of the digestive tract through which feces pass to the outside (60)

arteries   blood vessels that carry blood away from the heart (33)

atrium   an upper chamber of the heart (32)

autoimmune disease   a disease in which the immune system attacks the cells of the body it is meant to protect (136)

axon   a long cell fiber in the nervous system that transfers intercellular messages (77)

B

B cell   an immune-system cell that matures in bones and makes antibodies (131)

bile   a green liquid made by the liver and stored in the gallbladder; used in fat digestion (59)

blood   a connective tissue made up of platelets, white blood cells, red blood cells, and plasma (30)

blood pressure   the amount of force exerted by blood on the inside walls of a blood vessel (35)

brain   the mass of nerve tissue that is the main organ of the nervous system (79)

bronchi   (BRAHNG kie) the two tubes that connect the lungs with the trachea (41)

budding   a type of asexual reproduction in which a small part of the parent's body develops into an independent organism (102)

bulimia   a disorder characterized by binge eating followed by induced vomiting to rid the body of food (154)

C

Calorie   a unit that expresses the amount of energy found in food (148)

cancer   a condition in which certain body cells begin dividing at an uncontrolled rate (136)

capillaries   the smallest blood vessels (33)

carbohydrate   a biochemical composed of one or more simple sugars bonded together that is used to provide and store energy (149)

cardiac muscle   the type of muscle found in the heart (12)

cardiovascular system   a collection of organs that transport blood to and from your body's cells; the organs in this system include the heart, the arteries, and the veins (30)

cartilage   a flexible tissue that gives support and protection but is not rigid like bone (10)

cellular respiration   the process of producing ATP in the cell from oxygen and glucose; releases carbon dioxide and water (42)

central nervous system   a collection of organs that processes all incoming and outgoing messages from the nerves; the organs in this system include the brain and the spinal cord (76)

cerebellum   (SER uh BEL uhm) the part of the brain that keeps track of the body's position (80)

cerebrum   the part of the brain that detects touch, sight, sound, odor, taste, pain, heat, and cold and controls all voluntary acts, including thought (79)

chemical digestion   the process in which large molecules are broken down into simpler molecules or chemical building blocks (55)

cochlea   (KAHK lee uh) an ear organ that converts sound waves into electrical impulses (86)

compact bone   the type of bone tissue that does not have open spaces (9)

cones   photoreceptors that can detect bright light and help you see colors (84)

connective tissue   one of the four main types of tissue in the body; functions include support, protection, insulation, and nourishment (5)

D

dendrite   a short, branched extension of a neuron where the neuron receives impulses from other cells (77)

depressant   a drug that slows the actions of the central nervous system (155)

dermis the layer of skin below the epidermis (17)

diaphragm (DIE uh FRAM) the sheet of muscle underneath the lungs of mammals that helps draw air into the lungs (42)

digestive system a collection of organs that break down food so that it can be used by the body; the organs in this system include the stomach, the pancreas, the liver, the gallbladder, the small intestine, and the large intestine (54)

drug any chemical substance that causes a physical or emotional change in a person (155)

drug abuser a person who takes drugs for a purpose other than to relieve a medical condition (159)

E

egg a sex cell produced by a female (103)

embryo an organism in the earliest stage of development (110)

enamel the outermost layer of a tooth; the hardest material in the body (56)

endocrine system a collection of glands that control body-fluid balance, growth, and sexual development (88)

enzyme a protein that makes it possible for certain chemical reactions to occur quickly (55)

epidermis the outermost layer of the skin (17)

epididymis (EP uh DID i mis) the area of the testes where sperm are stored before they enter the vas deferens (106)

epithelial tissue one of the four main types of tissue in the body; the tissue that covers and protects underlying tissue (4)

esophagus (i SAWF uh guhs) a long, straight tube that connects the throat to the stomach (56)

excretion the process of removing wastes from the body; term used only when substances must pass through a membrane in order to leave the body (62)

extensor a muscle that straightens part of the body (13)

external fertilization the fertilization of eggs by sperm that occurs outside the body of the female (104)

F

fallopian tube the tube that leads from an ovary to the uterus (107)

farsighted describes someone who has better vision for distant objects than for near ones (85)

fat energy-storing nutrients that help the body store some vitamins (149)

feedback controls a system that turns endocrine glands on or off (90)

fetus an embryo during the later stages of development within the uterus (112)

flexor a muscle that bends part of the body (13)

fragmentation a type of reproduction in which an organism breaks into two or more parts, each of which may grow into a separate individual (102)

G

gallbladder a small, baglike organ that stores bile (59)

gland a group of cells that make special chemicals for the body (88)

H

hair follicle a small organ in the dermis layer of the skin that produces hair (18)

hallucinogen (huh LOO si nuh juhn) a drug that distorts the senses, causes changes in mood and thought processes, and causes hallucinations (158)

helper T cell an immune-system cell that activates killer T cells and B cells (132)

hemoglobin (HEE moh GLOH bin) the protein in red blood cells that attaches to oxygen so that oxygen can be carried through the body (30)

homeostasis (HOH mee OH STAY sis) the maintenance of a stable internal environment (4)

hormone a chemical messenger that carries information from one part of an organism to the other; in mammals, hormones are made by the endocrine glands (88)

hygiene methods of preserving and protecting your health (161)

I

immune system a collection of cells, tissues, and organs that fight disease-causing agents (131)

immunity resistance to a disease (128)

implantation the process in which an embryo imbeds itself in the lining of the uterus (110)

impulse an electrical message that passes along a neuron (77)

infectious disease a disease caused by a pathogen (126)

infertile the state of being unable to have children (109)

integumentary system (in TEG yoo MEN tuhr ee) a collection of organs that helps the body maintain a stable and healthy internal environment; the organs in this system include skin, hair, and nails (16)

internal fertilization the fertilization of an egg by sperm that occurs inside the body of a female (104)

iris the colored part of the eye (85)

J

joint the place where two or more bones connect (10)

K

kidney a bean-shaped organ that removes many harmful substances from the blood (63)

killer T cell an immune-system cell that kills body cells infected with pathogens (132)

L

large intestine a large organ that reabsorbs water from the digestive tract and stores, compacts, and eliminates indigestible material from the body (60)

larynx (LER ingks) the area of the throat that contains the vocal cords (41)

lens a curved, transparent object that forms an image by refracting light (85)

ligament a strong band of tissue that connects bones to bones (11)

liver a large, reddish brown organ that produces bile and stores nutrients; the liver has more than 200 functions in the body (59)

lung a saclike organ that takes oxygen from the air and delivers it to the blood (41)

lymph the fluid and particles absorbed into lymph capillaries (38)

lymph nodes small, bean-shaped organs that contain small fibers that work like nets to remove particles from the lymph (39)

lymphatic system a collection of organs that collect extracellular fluid and return it to the blood; the organs in this system include the lymph nodes and the lymphatic vessels (38)

M

macrophage (MAK roh FAYJ) an immune-system cell that engulfs pathogens (131)

malnutrition a disorder resulting from not consuming the right combination of nutrients (154)

marsupial a mammal that gives birth to live, partially developed young that continue to develop inside the mother's pouch or skin fold (105)

mechanical digestion the breaking, crushing, and mashing of food (55)

medulla (mi DOOL uh) the part of the brain that connects to the spinal cord and controls many involuntary processes in the body (80)

meiosis (mie OH sis) cell division that produces sex cells (103)

melanin a darkening chemical in the skin that determines skin color (16)

memory B cell an immune-system cell that "remembers" how to make a specialized antibody for a particular pathogen (134)

menstruation the monthly discharge of blood and tissue from the uterus (107)

mineral an element that is essential for good health (150)

monotreme a mammal that lays eggs (105)

motor neuron a neuron that sends impulses from the brain and spinal cord to other systems (78)

muscle tissue one of the four main types of tissue in the body; contains cells that contract and relax to produce movement (5)

muscular system a collection of organs whose primary function is movement; organs in this system include the muscles and the connective tissue that attaches them to bones (12)

N

narcotics drugs made from opium (159)

nearsighted describes someone who has better vision for near objects than for distant ones (85)

nephron a microscopic filter in the kidney that removes a variety of harmful substances from the blood (63)

nerve an axon bundled together with blood vessels and connective tissue (78)

nervous system a collection of organs that gather and interpret information about the body's internal and external environment and respond to that information; the organs in this system include the brain, nerves, and spinal cord (76)

nervous tissue one of the four main types of tissue in the body; the tissue that sends electrical signals through the body (4)

neuron a specialized cell that transfers messages throughout the body in the form of fast-moving electrical energy (77)

nicotine a chemical stimulant found in tobacco that increases heart rate and blood pressure (157)

noninfectious disease a disease that cannot spread from one person to another (126)

nutrient a substance that must be consumed or taken in by an organism to promote normal growth, maintenance, and repair (55, 148)

O

obesity a disorder characterized by an extremely high percentage of body fat (154)

optic nerve a nerve that transfers electrical impulses from the eye to the brain (84)

organ a combination of two or more tissues that work together to perform a specific function in the body (5)

organ system a group of organs that works together to perform body functions (5)

ovulation the process in which an egg is ejected through the ovary wall (107)

P

pancreas a fish-shaped organ between the stomach and small intestine that produces enzymes for chemical digestion (58)

pasteurization (PAS tuhr i ZAY shuhn) a method of heating food and beverages to kill bacteria (128)

pathogen an agent that causes a disease (126)

penis the male reproductive organ that transfers semen into the female's body during sexual intercourse (106)

peripheral nervous system the collection of communication pathways, or nerves, whose primary function is to transfer information from all areas of the body and the outside environment to the central nervous system and from the central nervous system to the rest of the body (76)

peristalsis (PER uh STAHL sis) a rhythmic muscle contraction in the digestive tract (56)

pharynx (FER ingks) the upper portion of the throat (41)

photoreceptors specialized neurons in the retina that detect light (84)

placenta a special organ of exchange that provides a developing fetus with nutrients and oxygen (111)

placental mammal a mammal that nourishes its unborn offspring with a placenta inside the uterus and gives birth to well-developed young (105)

plasma the fluid part of blood (30)

platelet a cell fragment that helps clot blood (31)

prescription a note written by a doctor to allow a patient to buy a medicine (156)

protein a biochemical that is composed of amino acids; its functions include regulating chemical reactions, transporting and storing materials, and providing support (149)

puberty the time of life when the sex organs become mature (106)

pulmonary circulation the circulation of blood between the heart and lungs (34)

pupil the opening to the inside of the eye (85)

R

receptor a specialized cell, sometimes a dendrite, that detects changes inside or outside the body (78)

red blood cell a cell that carries oxygen from the lungs to all cells of the body and carries carbon dioxide back to the lungs to be exhaled (30)

reflex a quick, involuntary response to a stimulus (82)

respiration the exchange of gases between living cells and their environment; includes breathing and cellular respiration (40); *see cellular respiration*

respiratory system a collection of organs whose primary function is to take in oxygen and expel carbon dioxide; the organs of this system include the lungs, the throat, and the passageways that lead to the lungs (40)

retina a layer of light-sensitive cells in the back of the eye (84)

rods photoreceptors that detect very dim light (84)

S

salivary glands organs located around the mouth that produce a liquid that begins chemical digestion (56)

saturated fat a type of fat found in meats, dairy products, coconut oil, and palm oil; known to raise blood cholesterol levels (150)

scrotum a skin-covered sac that hangs from the male body and contains the testes (106)

semen a mixture of sperm and fluids (106)

seminiferous tubules (SEM uh NIF uhr uhs TOO BYOOLZ) the coiled tubes inside the testes where sperm cells are produced (106)

sensory neuron a special neuron that gathers information about what is happening in and around the body and sends this information on to the central nervous system (78)

sexual reproduction reproduction in which two sex cells join to form a zygote; sexual reproduction produces offspring that share characteristics of both parents (103)

sexually transmitted disease a disease that can pass from an infected person to an uninfected person during sexual contact (109)

skeletal muscle the type of muscle that moves bones and helps protect inner organs (12)

skeletal system a collection of organs whose primary function is to support and protect the body; the organs in this system include bones, cartilage, ligaments, and tendons (8)

small intestine a muscular tube about 2.5 cm in diameter and up to 6 m long; the site of most chemical digestion (58)

smooth muscle the type of muscle found in the blood vessels and the digestive tract (12)

sperm a sex cell produced by a male (103)

spleen an organ that filters blood and produces lymphocytes (39)

spongy bone a type of bone tissue that has many open spaces and contains marrow (9)

stimulant a drug that speeds up the action of the central nervous system (155)

stomach a muscular, baglike organ of the digestive tract that is attached to the lower end of the esophagus (57)

stress a physical and mental response to situations that create pressure (163)

sweat glands small organs in the dermis layer of the skin that release sweat (16)

systemic circulation the circulation of blood between the heart and the body (excluding the lungs) (34)

T

T cell an immune-system cell that matures in the thymus (131)

tendon a tough connective tissue that connects skeletal muscles to bones (13)

testes the organs in the male reproductive system that make sperm and testosterone (106)

thymus a lymph organ that produces lymphocytes (39)

tissue a group of similar cells that work together to perform a specific job in the body (4)

tonsils small masses of soft tissue located at the back of the nasal cavity, on the inside of the throat, and at the back of the tongue (39)

trachea (TRAY kee uh) the air passageway from the larynx to the lungs (41)

U

umbilical cord the cord that connects an embryo to a placenta (111)

unsaturated fat a type of fat that usually comes from plant sources and helps reduce blood cholesterol levels (150)

ureter a slender tube that carries urine from each kidney to the urinary bladder (63)

urethra in males, a slender tube that carries urine and semen through the penis to the outside; in females, a slender tube that carries urine to the outside (63, 106)

urinary bladder a baglike organ that stores urine until it can be eliminated through the urethra (63)

urinary system a collection of organs that remove waste from the blood; this system includes the kidneys, ureters, urethra, and the urinary bladder (62)

urine a concentrated mixture of waste materials that forms in the nephrons of the kidney (63)

uterus the organ in the female reproductive system where a zygote grows and develops (107)

V

vaccine a substance that helps the body develop an immunity to a pathogen (128)

vagina the passageway in the female reproductive system that receives sperm during sexual intercourse (107)

vas deferens (vas DEF uh RENZ) a tube in males where sperm is mixed with fluids to make semen (106)

veins blood vessels that direct blood to the heart (33)

ventricle a lower chamber of the heart (32)

villi fingerlike projections on the inside wall of the small intestine (58)

vitamin an organic compound that controls many body functions, including cell growth and hormone production (151)

W

white blood cell a blood cell that protects the body against pathogens (31)

Z

zygote a fertilized egg (103)

Index

Credits

Abbreviations used: (t) top, (c) center, (b) bottom, (l) left, (r) right, (bkgd) background

ILLUSTRATIONS

All illustrations, unless otherwise noted below by Holt, Rinehart and Winston.

Table of Contents vi(tl), Kip Carter; vii(tr), Keith Kasnot; vii(br), Morgan-Cain Associates.

Chapter One Page 4 (c,cl), Morgan-Cain & Associates; 5 (cl), Morgan-Cain & Associates; 5 (c), Morgan-Cain & Associates; 5 (tr), Morgan-Cain & Associates; 6, Christy Krames; 7, Christy Krames; 9, Keith Kasnot; 10, John Huxtable/Black Creative; 11, John Huxtable/Black Creative; 13 (br), Christy Krames; 17 (br), Morgan-Cain & Associates; 17 (cr), Marty Roper/Planet Rep; 19 (t), Morgan-Cain & Associates; 22 (br), John Huxtable/Black Creative; 24 (br), Christy Krames; 25 (tr), Morgan-Cain & Associates.

Chapter Two Page 30 (tr), Christy Krames; 31 (b), Keith Kasnot; 32, Kip Carter; 33 (c), Kip Carter; 34 (b), Kip Carter; 36 (tl), Jared Schneidman Design; 36 (br), Marty Roper/Planet Rep; 38 (br), Kip Carter; 39 (tr), Christy Krames; 40 (bl), Christy Krames; 41 (b), Christy Krames; 42 (br), Christy Krames; 42 (bc,bl), Kip Carter; 46 (br), Christy Krames; 48 (tr), Kip Carter; 49 (cr), Kip Carter.

Chapter Three Page 54 (bc), Christy Krames; 55 (b), Brian Evans; 56 (tl,bl), Keith Kasnot; 57 (cl), Christy Krames; 57 (cr), Brian Evans; 58 (tl), Marty Roper/Planet Rep; 58 (cl), Christy Krames; 58 (c), Brian Evans; 59, Christy Krames; 60 (tl), Christy Krames; 62 (cl), Christy Krames; 63 (cr), Keith Kasnot; 68 (cr), Brian Evans; 68 (tr), Keith Kasnot; 69 (tr), Christy Krames; 70 (br), Keith Kasnot; 70 (tr), Brian Evans.

Chapter Four Page 76 (bl), Christy Krames; 77 (b), Scott Thorn Barrows/The Neis Group; 78 (b), Scott Thorn Barrows/The Neis Group; 79 (bc), Brian Evans; 80 (b), Brian Evans; 81 (tr), Christy Krames; 83 (b), Morgan-Cain & Associates; 84 (bc), Keith Kasnot; 84 (bl), Carlyn Iverson; 85 (t), Keith Kasnot; 86 (b), Christy Krames; 87 (tr), Keith Kasnot; 88 (b), Dan McGeehan/Koralick Associates; 89 (b), Christy Krames; 90, Christy Krames; 94 (br), Keith Kasnot; 95 (cl), Dan McGeehan/Koralick Associates; 96 (tr), Christy Krames; 97 (tr), Christy Krames.

Chapter Five Page 103 (bl), Rob Schuster/Hankins and Tegenborg; 106 (bl), Keith Kasnot; 107 (tr), Keith Kasnot; 108 (br), Rob Schuster/Hankins and Tegenborg; 110 (cl), David Fischer; 111 (cr), Christy Krames; 121 (cr), Sidney Jablonski; 122 (bl), Morgan-Cain & Associates.

Chapter Six Page 130 (br), Scott Thorn Barrows/The Neis Group; 131 (bl), Scott Thorn Barrows/The Neis Group; 132-133, Blake Thornton/Rita Marie; 134 (l), Stephen Durke/Washington Artists; 140 (br), Blake Thornton/Rita Marie; 142 (bc), Scott Thorn Barrows/The Neis Group; 143 (tr), Sidney Jablonski.

LabBook Page 178, Morgan-Cain Associates;

Chapter Seven Page 159 (br), Marty Roper/Planet Rep; 164 (c), Uhl Studios, Inc.; 169 (tl), Marty Roper/Planet Rep.

Appendix 188 (t), Terry Guyer; 192 (b), Mark Mille/Sharon Langley; 196-197, Kristy Sprott.

PHOTOGRAPHY

Cover: VCG/FPG International

Title page: (cr), Frans Lanting/Minden Pictures; (tc), Kim Taylor/Bruce Coleman, Inc.

Feature Borders: Unless otherwise noted below, all images copyright ©2001 PhotoDisc/HRW. "Across the Sciences" 72, 122, all images by HRW; "Careers" 144, sand bkgd and Saturn, Corbis Images; DNA, Morgan Cain & Associates; scuba gear, ©1997 Radlund & Associates for Artville; "Eureka" 27, 99 ©2001 PhotoDisc/HRW; "Health Watch" 51, 73, 145, 173, dumbbell, Sam Dudgeon/HRW Photo; aloe vera, EKG, Victoria Smith/HRW Photo; basketball, ©1997 Radlund & Associates for Artville; shoes, bubbles, Greg Geisler; "Scientific Debate" 95, Sam Dudgeon/HRW Photo; "Weird Science" 50, mite, David Burder/Stone; atom balls, J/B Woolsey Associates; walking stick, turtle, EclectiCollection.

Table of Contents: iv(tl), Lennart Nilsson/Albert Bonniers Forlag AB, A CHILD IS BORN; iv(cl), Tektoff-RM/CNRI/Science Photo Library/Photo Researchers, Inc.; iv(bl), Chris Hamilton; v(tr), K. H. Kjeldsen/Science Photo Library/Photo Researchers, Inc.; v(tcr), John Huxtable/Black Creative; v(cr), SuperStock; v(bc), Sam Dudgeon/HRW Photo; vi(cl), Prof. P. Motta/Department of Anatomy/University "La Sapienza" Rome/Science Photo Library/Photo Researchers, Inc.; vi(b), Sam Dudgeon/HRW Photo; vii(cr), Sam Dudgeon/HRW Photo.

Chapter One: 2-3, AFP/Corbis; 3, HRW Photo; 4-5(b), David Madison/Stone; 10(tl), Peter Dazeley/Stone; 10(bc, br, bl), SP/FOCA/HRW Photo; 12(bc), Bob Torrez/Stone; 12(bl), Dr. E.R. Degginger; 12(br), Manfred Kage/Peter Arnold, Inc.; 12(cl), G.W. Willis/BPS/Stone; 14(bl), Chris Hamilton; 15 Shelby Thorner/David Madison, 18(c), Dr. Robert Becker/Custom Medical Stock Photo; 19(cr), Dr. P. Marazzi/Science Photo Library/Photo Researchers, Inc.; 23 Peter Dazeley/Stone; 26(tr), Dan McCoy/Rainbow; 27(cl), Gamma-Liaison; 27(tr), Huntsville Times.

Chapter Two: 28-29, Bruce Iverson; 29, HRW Photo; 30(bl), Dr. Dennis Kunkel/Phototake NYC; 31, Don Fawcett/Photo Researchers, Inc.; 33 (c), O. Meckes/Nicole Ottawa/Photo Researchers, Inc.; 33(cr), David Phillips/Science Source/Photo Researchers, Inc.; 35(tr), Custom Medical Stock Photo; 35(br), James Wilson/Woodfin Camp & Associates; 37(tr), Ken Wagner/Phototake NYC; 43(cl, cr), Matt Meadows/Peter Arnold, Inc.; 46(c), Dr. Dennis Kunket/Phototake; 47(tr), Don Fawcett/Photo Researchers, Inc.; 49(bl), Dr. Dennis Kunkel/Phototake NYC; 50(tr), Index Stock Photography; 51(tr), Russell Dian/HRW Photo; 51(bl), Jim Gripe/Pivot Media.

Chapter Three: 52-53, Bod Daemmrich; 53, HRW Photo; 61(br), Prof. P. Motta/Dept. of Anatomy/University "La Sapienza" Rome/Science Photo Library/Photo Researchers, Inc.; 61(tr), The Stock Market; 64(br), Image Bank; 65(tr), Stephen J. Krasemann/DRK Photo; 65(cr), E.K. Martin and Associates; 66, Sam Dudgeon/HRW Photo; 73 J. H. Robinson/Photo Researchers, Inc.

Chapter Four: 74-75, Omikron/Photo Researchers, Inc.; 75, HRW Photo; 85 Bruno Joachim/Liaison; 87 Louis Psihoyos/Matrix; 91 Will & Deni McIntyre/Photo Researchers, Inc.; 99 Journal of Nuclear Medicine.

Chapter Five: 100-101, Lennart Nilsson; 101, HRW Photo; 102(bl), Visuals Unlimited/Cabisco; 102(br), Innerspace Visions; 104(tl), Michael Fogden/Animals Animals; 104(bl), Photo Researchers, Inc.; 104(cr), Guy Mannering/Bruce Coleman; 105 Dr. E. R. Degginger/Bruce Coleman; 108(tl), Chip Henderson/Stone; 110(b), Lennart Nilsson; 111 Petit Format/Nestle/Science Source/Photo Researchers, Inc.; 112 Lennart Nilsson; 113(tr), Lennart Nilsson/Albert Bonniers Forlag AB, A CHILD IS BORN; 113(cr), Keith/Custom Medical Stock Photo; 115 NASA; 117, Victoria Smith/HRW Photo; 118 Guy Mannering/Bruce Coleman, Inc.; 120 Lennart Nilsson/Albert Bonniers Forlag AB, BEING BORN; 123(inset), Vince Viverito, Jr./Richard Wolf Medical Instruments Corp., Vernon Hills, IL; 123(cr), Tom McCarthy/Rainbow.

Chapter Six: 124-125, Oliver Meckes/Photo Researchers; 125, HRW Photo; 126(br), CNRI/Science Photo Library/Photo Researchers, Inc.; 126(bc), Tektoff-RM/CNRI/Science Photo Library/Photo Researchers, Inc.; 126(bl), Manfred Kage/Peter Arnold; 127 Kent Wood/Photo Researchers, Inc.; 135(tr), Visuals Unlimited/George Musil; 135(bc), Image Copyright ©<2000>PhotoDisc, Inc.; 135(cr), K. H. Kjeldsen/Science Photo Library/Photo Researchers, Inc.; 135(cl), SuperStock; 136(tl), Clinical Radiology Dept., Salisbury District Hospital/Science Photo Library/Photo Researchers, Inc.; 136(bl, br), Dr. A. Liepins/Science Photo Library/Photo Researchers, Inc.; 137 Lennart Nilsson; 141 Dr. A. Liepins/Science Photo Library/Photo Researchers, Inc.; 144(all), Chris Mooney/HRW Photo; 145 E. R. Degginger/Bruce Coleman.

Chapter Seven: 146-147, Arthur Tilley/FPG International; 147, HRW Photo; 152 John Kelly/Stone; 157(bl), E. Dirksen/Photo Researchers, Inc.; 157(tr), ©<2000>Stephen Foster; 157(br), Dr. Andrew P. Evans/Indiana University; 158 Spencer Grant/Photo Researchers, Inc.; 162 Rob Van Petten/Image Bank; 163 Wally McNamee/Corbis; 166, Peter Van Steen/HRW Photo172 Manfred Kage/Peter Arnold; 173 SuperStock.

LabBook: "LabBook Header": "L", Corbis Images, "a", Letraset-Phototone, "b" and "B", HRW, "o" and "k", Images Copyright ©<2000>PhotoDisc, Inc. 175(cl), Michelle Bridwell/HRW Photo; 175(br), Image Copyright ©<2000>Photodisc, Inc.; 176(bl), Stephanie Morris/HRW Photo; 177(tr), Jana Birchum/HRW Photo.

Appendix: 200 CENCO.

Sam Dudgeon/HRW Photos: all Systems of the Body background photos, p. viii-1, 65 (bl), 6, 8, 13(tr, tc), 14(c), 16, 21, 22(c), 44, 54-55(t), 66, 67, 80, 82, 89, 90, 92, 93, 94, 119, 161, 169(cr), 174, 175(bc), 176(br, tl), 176(tr), 177(tl), 180, 181, 184, 189(br).

Peter Van Steen/HRW Photos: p. 11(all), 18(l, br), 72(cl), 114(all), 128, 130, 135(cl), 140, 148, 149(all), 150, 153, 155, 156, 157(cr), 162(br), 163(br), 165(all), 168, 170, 171(tr, cr), 177(b), 189(tr).

John Langford/HRW Photos: p. 129, 175(tr).

Acknowledgements continued from page iii.

Alyson Mike
Science Teacher
East Valley Middle School
East Helena, Montana

Donna Norwood
Science Teacher and Dept. Chair
Monroe Middle School
Charlotte, North Carolina

James B. Pulley
Former Science Teacher
Liberty High School
Liberty, Missouri

Terry J. Rakes
Science Teacher
Elmwood Junior High School
Rogers, Arkansas

Elizabeth Rustad
Science Teacher
Crane Middle School
Yuma, Arizona

Debra A. Sampson
Science Teacher
Booker T. Washington Middle School
Elgin, Texas

Charles Schindler
Curriculum Advisor
San Bernadino City Unified Schools
San Bernadino, California

Bert J. Sherwood
Science Teacher
Socorro Middle School
El Paso, Texas

Patricia McFarlane Soto
Science Teacher and Dept. Chair
G. W. Carver Middle School
Miami, Florida

David M. Sparks
Science Teacher
Redwater Junior High School
Redwater, Texas

Elizabeth Truax
Science Teacher
Lewiston-Porter Central School
Lewiston, New York

Ivora Washington
Science Teacher and Dept. Chair
Hyattsville Middle School
Washington, D.C.

Elsie N. Waynes
Science Teacher and Dept. Chair
R. H. Terrell Junior High School
Washington, D.C.

Nancy Wesorick
Science and Math Teacher
Sunset Middle School
Longmont, Colorado

Alexis S. Wright
Middle School Science Coordinator
Rye Country Day School
Rye, New York

John Zambo
Science Teacher
E. Ustach Middle School
Modesto, California

Gordon Zibelman
Science Teacher
Drexel Hill Middle School
Drexell Hill, Pennsylvania

Self-Check Answers

Chapter 1—Body Organization and Structure

Page 14: Curl-ups use flexor muscles; push-ups use extensor muscles.

Page 17: Blood vessels belong to the cardio-vascular system.

Chapter 2—Circulation and Respiration

Page 34: The hollow tube shape of arteries and veins allows blood to reach all parts of the body. Valves in the veins prevent blood from flowing backward.

Page 38: Like blood vessels, lymph capillaries receive fluid from the spaces surrounding cells. The fluid absorbed by lymph capillaries flows into lymph vessels. These vessels drain into large neck veins instead of into an organ, such as the heart. Lymph does not deliver oxygen and nutrients.

Chapter 3—The Digestive and Urinary Systems

Page 59: Bile is involved in the physical digestion because emulsification does not change the chemical composition of the fat molecules; it only increases the surface area of each fat droplet.

Chapter 4—Communication and Control

Page 81: 1. cerebrum 2. cerebellum 3. to protect the spinal cord

Chapter 5—Reproduction and Development

Page 103: In asexual reproduction, one animal produces offspring that are genetically identical to itself. In sexual reproduction, the genes of two individuals are mixed when sex cells join to form a zygote. This zygote develops into a unique individual.

Page 111: The uterus provides the nutrients and protection that the embryo needs to continue growing. The uterus is also the only place the placenta will form.

Chapter 6—Body Defenses and Disease

Page 127: If Jackie did not wash the counter after cutting up the meat, bacteria could grow on the counter where the meat was. This bacteria could contaminate her brother's sandwich.

Chapter 7—Staying Healthy

Page 153: You have eaten two servings from the bread, cereal, rice, pasta group; one serving from the fruit group, and one from the milk, yogurt and cheese group.

Page 156: Regular use of some drugs may cause tolerance or addiction. Withdrawal symptoms occur when the body does not receive a drug that it is addicted to.

Page 159: 1. A hallucination is a vision or sound that is not real. 2. Heroin is highly addictive and potentially deadly. If shared needles are used to inject it, users risk getting diseases like hepatitis or AIDS.